Structural Design and Properties of Coordination Polymers

Special Issue Editor
George E. Kostakis

MDPI • Basel • Beijing • Wuhan • Barcelona • Belgrade

MDPI

Special Issue Editor
George E. Kostakis
University of Sussex
UK

Editorial Office
MDPI
St. Alban-Anlage 66
Basel, Switzerland

This edition is a reprint of the Special Issue published online in the open access journal *Crystals* (ISSN 2073-4352) in 2017 (available at: http://www.mdpi.com/journal/crystals/special_issues/ Coordination_Polymers).

For citation purposes, cite each article independently as indicated on the article page online and as indicated below:

Lastname, F.M.; Lastname, F.M. Article title. *Journal Name* **Year**, *Article number*, page range.

First Edition 2018

ISBN 978-3-03842-801-5 (Pbk)
ISBN 978-3-03842-802-2 (PDF)

Table of Contents

About the Special Issue Editor

George E. Kostakis, Ph. D., studied Chemistry at the University of Ioannina (Ioannina, Greece) and became familiar with key areas of organic chemistry and coordination chemistry. During his Ph.D., he worked as a Marie Curie Fellow with Casella (Italy) and Nordlander (Sweden). In December 2007, he was employed as post doc at the Inorganic Chemistry Institute, Karlsruhe, Germany, under the supervision of Prof. Annie K. Powell and subsequently worked as Senior Researcher at the Institute of Nanotechnology. He joined the School of Life Sciences, University of Sussex, in September 2013 and focused his research interests on ground-breaking aspects of coordination chemistry, including a) Exploration of the catalytic properties of polynuclear 3d/4f Coordination Clusters (CCs), b) Synthesis of new coordination polymers based on benzotriazole ligands and exploration of their catalytic activity and c) Development of the Polynuclear Inorganic Clusters Database (PICD).

Preface to "Structural Design and Properties of Coordination Polymers"

Coordination polymers (CPs) or metalorganic frameworks (MOFs), have become one of the most prominent branches of inorganic and materials chemistry due to the extensive variety of their applications. This book covers a range of recent developments in coordination polymer chemistry. Wang and Chuang et al. report a reversible Single-Crystal-to-Single-Crystal structural transformation and sorption studies (Crystals 2017, 7(12), 364; doi:10.3390/cryst7120364). Sun et al. report a three-dimensional porous CP, that showed the selective adsorption of CO2 over CH4 (Crystals 2017, 7(12), 370; doi:10.3390/cryst7120370). McCormick et al. describe the solvent-dependent disorder in a series of CPs, demonstrating the influence of CP synthesis, even when the overall topology of the framework is not affected (Crystals 2018, 8(1), 6; doi:10.3390/cryst8010006). Bai et al. report the synthesis of a one-dimensional CP containing cyclic [Ag4] clusters supported by a hybrid pyridine and thioether functionalized 1,2,3-triazole (Crystals 2018, 8(1), 16; doi:10.3390/cryst8010016). Zhang and Hu et al. describe Prussian Blue analogue mesoframes for enhanced aqueous sodium-ion storage (Crystals 2018, 8(1), 23; doi:10.3390/cryst8010023). Janiak et al. incorporate the thiazolo[5,4-d]thiazole unit into a CP with interdigitated structure and reported for the first time the sorption characteristics of a tztz-functionalized porous MOF material (Crystals 2018, 8(1), 30; doi:10.3390/cryst8010030). Ghosh et al. report a Zn(II)-based MOF built from a neutral N-donor linker and SiF62 anions that exhibits a two-step structural transformation, maintaining the crystallinity of the framework (Crystals 2018, 8(1), 37; doi:10.3390/cryst8010037). Kostakis et al. describe a 12-fold interpenetrated CP with ths topology built from Cu(II) ions and a glycine-based pseudopeptidic ligand and its use as a heterogeneous catalyst in a A3 coupling reaction (Crystals 2018, 8(1), 47; doi:10.3390/cryst8010047). Ingram incorporates anthracene-based ligands and Lanthanide ions to synthesize MOFs and studied their photoluminescence and radioluminescence properties (Crystals 2018, 8(1), 53; doi:10.3390/cryst8010053). Gu and Kirillov et al. discuss the suitability of multifunctional aromatic carboxylic acids as versatile building blocks for the hydrothermal design of CPs (Crystals 2018, 8(2), 83; doi:10.3390/cryst8020083).

George E. Kostakis

Special Issue Editor

crystals

MDPI

Article

Reversible Single-Crystal-to-Single-Crystal Structural Transformation in a Mixed-Ligand 2D Layered Metal-Organic Framework: Structural Characterization and Sorption Study

Chih-Chieh Wang [1],*, Szu-Yu Ke [1], Kuan-Ting Chen [1], Yi-Fang Hsieh [1], Tzu-Heng Wang [1], Gene-Hsiang Lee [2] and Yu-Chun Chuang [3],*

[1] Department of Chemistry, Soochow University, Taipei 11102, Taiwan; sarah6017@yahoo.com.tw (S.-Y.K.); 04133019@scu.edu.tw (K.-T.C.); fangfangmaimai@gmail.com (Y.-F.H.); 06333002@scu.edu.tw (T.-H.W.)
[2] Instrumentation Center, National Taiwan University, Taipei 10617, Taiwan; ghlee@ntu.edu.tw
[3] National Synchrotron Radiation Research Center, Hsinchu 30076, Taiwan
* Correspondence: ccwang@scu.edu.tw (C.-C.W.); chuang.yc@nsrrc.org.tw (Y.-C.C.);
Tel.: +886-2-2881-9471 (ext. 6828) (C.-C.W.)

Academic Editor: George E. Kostakis
Received: 20 November 2017; Accepted: 5 December 2017; Published: 7 December 2017

Abstract: A 3D supramolecular network, $[Cd(bipy)(C_4O_4)(H_2O)_2]\cdot 3H_2O$ (**1**) (bipy = 4,4'-bipyridine and $C_4O_4^{2-}$ = dianion of $H_2C_4O_4$), constructed by mixed-ligand two-dimensional (2D) metal-organic frameworks (MOFs) has been reported and structurally determined by the single-crystal X-ray diffraction method and characterized by other physicochemical methods. In **1**, the $C_4O_4^{2-}$ and bipy both act as bridging ligands connecting the Cd(II) ions to form a 2D layered MOF, which are then extended to a 3D supramolecular network via the mutually parallel and interpenetrating arrangements among the 2D-layered MOFs. Compound **1** shows a two-step dehydration process with weight losses of 11.0% and 7.3%, corresponding to the weight-loss of three guest and two coordinated water molecules, respectively, and exhibits an interesting reversible single-crystal-to-single-crystal (SCSC) structural transformation upon de-hydration and re-hydration for guest water molecules. The SCSC structural transformation have been demonstrated and monitored by single-crystal and X-ray powder diffraction, and thermogravimetic analysis studies.

Keywords: coordination polymer; metal-organic framework; SCSC structural transformation; hydrogen bond; gas sorption

1. Introduction

Porous materials based on coordination polymers (CPs), or metal-organic frameworks (MOFs) [1,2] containing guest molecules are very attractive research field, not only owing to their designable structure, unusual flexibilities, but also on their tunable functional application [3–12]. In this field, dynamic porosity has been paid much attention on the structural characteristics, as well as on the function of guest inclusions [13–25]. In order to increase the flexibility of the porosity for guest inclusion, the design of host framework with a flexible skeleton is required. Based on this requirement, the dimensionality of host framework and binding forces for the component skeleton becomes an important target. A variety of supramolecular architectures have been built up with a large range of bonding forces, depending on the system, the interactions can range from M–L coordination bonds, to strong halogen [26–32] or hydrogen bonds [31,32], to much weaker forces, such as weak hydrogen bonds [33,34] and $\pi-\pi$ stacking of small aromatics [35–39]. N,N-based ligands, such as 4,4'-bipyridine (bipy) and pyrazine, have been widely used on the construction of many MOFs, including diamondoid, honeycomb, grid, T-shape, ladder, brick-wall, and octahedral frameworks [40–48]. In the previous study, a series of interpenetrated metal-organic coordination complexes, {[M(C_4O_4)(bipy)(H_2O)_2]·3H_2O}_\infty (M = Mn, Fe, Co, Ni; $C_4O_4{}^{2-}$ = dianion of $H_2C_4O_4$), have been synthesized under hydrothermal condition. The reversible de-/rehydration processes of guest water molecules in the 1D channels by heating or by vacuum were accompanied with color-changing and structural variation of the materials, which have been identified by MS measurements, UV spectrometry, and X-ray powder diffraction methods [49–51]. In our previous study [52], an isostructural coordination polymer, {[Cd(C_4O_4)(bipy)(H_2O)_2]·3H_2O}_n (**1**), has been synthesized under hydrothermal conditions and its reversible de-/rehydration property of guest water molecules in channels accompanying structural variation is demonstrated by thermogravimetric (TG) analysis, as well as temperature-dependent power X-ray diffraction. However, the detail structural information on the single-crystal-to-single-crystal (SCSC) transformation associated with the de-/rehydration of the guest water molecules in the channels, the second stage water de-/adsorption behavior of the coordinated water molecules, and the gas sorption property of **1** are interesting and worthy of further study. With our continuous effort on the structural transformation study of the two-step water de-/adsorption behavior in **1**, we report here on the exploration of the relationship between the structures and de-/rehydration behavior. The dehydrated species {[Cd(C_4O_4)(bipy)(H_2O)_2]}_n (**1a**), after heating **1** at 90 °C, show CO_2 gas sorption uptakes. Compound **1a** also shows remarkable reversibility to give rehydrated {[Cd(C_4O_4)(bipy)(H_2O)_2]·3H_2O}_n (**1b**), which is the same to parent **1**, when exposed to water vapor.

2. Results and Discussion

In the study, compound **1** was synthesized by solution method instead of hydrothermal method reported in previous study [52]. The reaction yield is obviously improved from 23.4% under hydrothermal condition to 70.4% by the solution method. The most relevant IR features are those associated with the chelating squarate ligands. Strong and broad absorptions occurring in the range of 1603–1323 cm^{-1} centered at 1470 cm^{-1} for **1** are characteristic of $C_4O_4{}^{2-}$ ions [53], which can be assigned to the vibrational modes representing mixtures of C–O and C–C stretching motions.

2.1. Structural Description of 1

The crystal structure of **1**, synthesized by the solution method, is re-determined and is almost the same as the crystal structure described in our previous report [52], with a 3D microporous supramolecular network being constructed by 2D-layered MOFs. The related bond lengths and angles around the Cd(II) ion are listed in Table 1. The squarate and bipy both act as a bridging ligand with $\mu_{1,3}$-*bis*-monodentate and *bis*-monodentate coordination mode, respectively, leading the formation of a 2D layered MOF (Figure 1b) along the *b* axis. The 2D MOF of **1** can be viewed in a simplified way using TOPOS [54,55] as a four-connected uninodal net with the point symmetry (Schläfli symbol) {$4^4.6^2$}. Adjacent 2D layers are arranged in mutually parallel and interpenetrated manners to complete its 3D supramolecular network, which generates 1D channels along the *c* axis intercalated with guest water molecules (Figure 1c). Hydrogen bonding interaction plays an important role on the stabilization of the 3D supramolecular network. In the crystal packing, the coordinated water molecules (O(3)) are held together with uncoordinated oxygen atoms (O(2)) of squarate ($C_4O_4{}^{2-}$) by means of intramolecular O–H···O hydrogen bonds with O···O distances of 2.664(5) and 2.698(5). The 3D supramolecular architecture is further reinforced by the O–H···O hydrogen bonds among the guest water molecules (O(4) and O(5)) and squarate ligands. Related bond distances and angles of O–H···O hydrogen bonds are listed in Table 2.

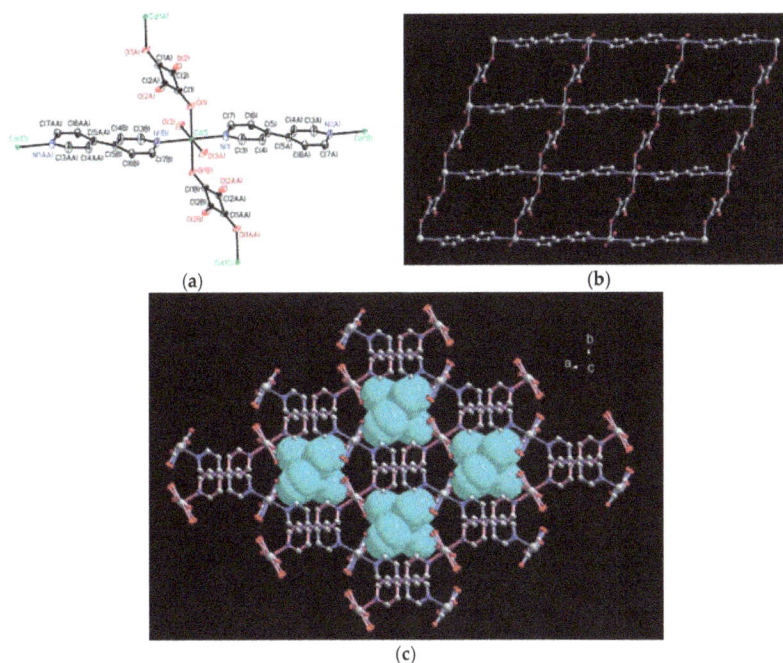

(a)

(b)

(c)

Figure 1. (a) Coordination environment of the Cd(II) ion in **1** with atom labelling scheme (ORTEP drawing, 50% thermal ellipsoids). The guest water molecules and H atoms are omitted for clarity. (b) The 2D MOF in **1** via the bridges of squarate and 4,4′-bipyridine; (c) The 3D supramolecular network with the 1D channels intercalated with guest water molecules (space-filling mode) viewing along the c axis.

Table 1. Bond lengths (Å) and angles (°) around Zn(II) ion in **1**, **1a**, and **1b**. [1]

Compound	1	1a	1b
Cd(1)–O(3)	2.264(3)	2.303(3)	2.265(2)
Cd(1)–O(3)$_i$	2.264(3)	2.303(3)	2.265(2)
Cd(1)–O(1)	2.294(3)	2.273(2)	2.296(2)
Cd(1)–O(1)$_i$	2.294(3)	2.273(2)	2.296(2)
Cd(1)–N(1)	2.341(4)	2.339(3)	2.340(2)
Cd(1)–N(1)$_i$	2.341(4)	2.339(3)	2.340(2)
O(3)–Cd(1)–O(3)$_i$	180	180.0	180.0
O(3)$_i$–Cd(1)–O(1)$_i$	92.77(12)	93.61(11)	92.76(7)
O(3)–Cd(1)–O(1)$_i$	87.23(12)	86.39(11)	87.24(7)
O(3)$_i$–Cd(1)–O(1)	87.23(12)	86.39(11)	87.24(7)
O(3)–Cd(1)–O(1)	92.77(12)	93.61(11)	92.76(7)
O(1)–Cd(1)–O(1)$_i$	180	180.0	180.0
O(3)$_i$–Cd(1)–N(1)	87.82(13)	91.88(11)	87.93(8)
O(3)–Cd(1)–N(1)	92.18(13)	88.12(11)	92.07(8)
O(1)$_i$–Cd(1)–N(1)	85.78(14)	86.46(11)	85.60(8)
O(1)–Cd(1)–N(1)	94.22(14)	93.54(11)	94.40(8)
O(3)$_i$–Cd(1)–N(1)$_i$	92.18(13)	88.12(11)	92.07(8)
O(3)–Cd(1)–N(1)$_i$	87.82(13)	91.88(11)	87.93(8)
O(1)$_i$–Cd(1)–N(1)$_i$	94.22(14)	93.54(11)	94.40(8)
O(1)–Cd(1)–N(1)$_i$	85.78(14)	86.46(11)	85.60(8)
N(1)–Cd(1)–N(1)$_i$	180.0	180.0	180.0

[1] Symmetry transformations used to generate equivalent atoms: $i = -x + 1/2, -y + 1/2, -z + 1$.

Table 2. The O–H···O hydrogen bonds for **1** and **1b**, respectively. [1]

	Compound 1			
D–H···A	D–H (Å)	H···A (Å)	D···A (Å)	∠D–H···A (°)
O(3)–H(3A)···O(2)$_i$	0.81(7)	1.88(7)	2.664(5)	162(6)
O(3)–H(3B)···O(2)$_{ii}$	0.87(7)	1.85(7)	2.698(5)	163(6)
O(4)–H(4A)···O(5)	0.83(5)	1.96(5)	2.762(5)	162(6)
O(5)–H(5A)···O(1)	0.71(8)	2.11(8)	2.807(5)	171(8)
O(5)–H(5B)···O(4)$_{iii}$	0.79(7)	2.00(7)	2.782(5)	168(7)
	Compound 1b			
D–H···A	D–H (Å)	H···A (Å)	D···A (Å)	∠D–H···A (°)
O(3)–H(3A)···O(2)$_i$	0.79(4)	1.89(4)	2.663(3)	166(4)
O(3)–H(3B)···O(2)$_{ii}$	0.81(4)	1.91(4)	2.698(3)	165(3)
O(4)–H(4A)···O(5)	0.82(3)	1.95(3)	2.766(3)	169(4)
O(5)–H(5A)···O(1)	0.81(4)	2.01(4)	2.808(3)	171(4)
O(5)–H(5B)···O(4)$_{iii}$	0.76(4)	2.04(4)	2.783(3)	168(4)

[1] Symmetry transformations used to generate equivalent atoms: $i = -x + 1/2, -y + 1/2, -z$; $ii = -x + 1/2, y - 1/2, -z + 1/2$; $iii = -x + 1, -y + 1, -z + 1$.

2.2. Water Adsorption Property of (1) by Cyclic TG Aanalsis and PXRD Measurements

In our previously thermal-stability study [52], the weight losses of the first and second steps were 11.1% and 7.7%, corresponding to the release of three guest water molecules (calc. 11.5%) and two coordinated water molecules (calc. 7.7%), respectively, and the first weight-loss step has been proven to be a reversible de-/adsorption of three guest water molecules by cyclic TG measurements under water vapor, as demonstrated by the display of de-/rehydration processes as a function of time and temperature [52]. This result can be correlated to the polarity of the pore surfaces, as the framework is adorned with high affinity for H_2O due to the presence of hydrophilic sites of $C_4O_4^{2-}$ ligands. Unlike the complete reversibility of guest water molecules in the first stage, the removal of coordinated water molecules is more likely to be only partially reversible in the second stage (Figure 2). When the guest and coordinated water molecules are completely removed with the weight loss of 18.8% after the temperature reaching 170 °C, only partial weight increase of 2.7% under water vapor has been recovered during the cooling process. Such heating and cooling processes have been repeated for five cycles (Figure 2) with weight-increasing and weight-decreasing percentages in the range of 2.7–3.2% to demonstrate the stable but incomplete water ad-/desorption behavior during the thermal re-/dehydration processes. This result indicates that, during the de-hydration period, the 3D supramolecular network and the coordination environment of Cd(II) ions of completely dehydrated form **1** might be changed after the removal of coordinated water molecules and may not be easily recovered to the original structure, which will be discussed in the next powder X-ray diffraction part.

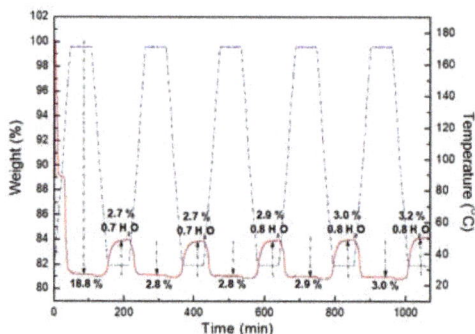

Figure 2. The cyclic TG measurements of water de-/adsorption processes at 170 °C and back to RT. Red solid line: the variation of weight loss with time; blue dashed line: the variation of temperature with time.

According to the TGA and single crystal diffraction results, the first structure change are dominated from packing guest water molecules loss and the second step is triggered by coordinated water molecules loss. To confirm the water re-absorption behavior, an in situ powder X-ray diffraction experiment was performed and shown in Figure 3. The sample in glass capillary was heated up to 90 °C initially and stays 5 min to remove the packing water molecules. Compare to the simulation pattern of rehydrated **1a** at 340 K, the diffraction pattern show the first step packing guest water molecules loss has been happened. Then the sample was taken into water and re-measured powder pattern at 30 °C. Clearly, the structure can be recovered effectively. However, it was failed to recover the structure while the coordinated water molecules be removed. A very large structure change occurred at the second water molecules loss step, which leads to an irreversible reaction.

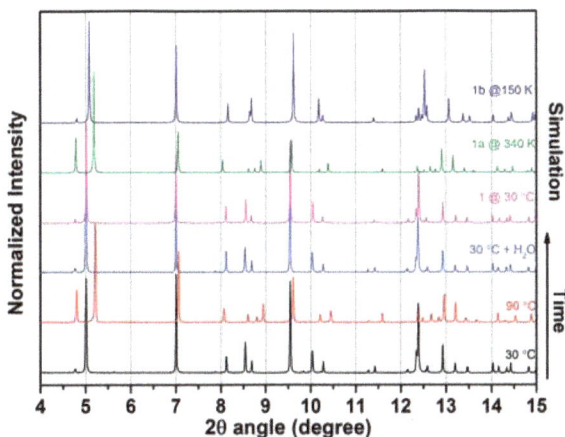

Figure 3. PXRD patterns of as synthesized **1** at RT (black), dehydrated **1a** at 90 °C (red), rehydrated **1b** at RT (blue), simulated **1** from single-crystal data (pink), simulated **1a** from single-crystal data at 90 °C (green), and simulated **1b** from single-crystal data at RT (deep blue).

2.3. SCSC Transformation Associated with Guest Water De-/Rehydration

According to the TG analysis and in situ PXRD measurements, the guest water molecules located into the channels can be removed from the host framework by heating up to 90 °C. The 3D supramolecular architecture of dehydrated **1** is rigid and stable without the guest water molecules, and the de-/adsorption processes of guest water molecules are reversible. Moreover, the thermal stability of the framework provides the opportunity to determine the crystal structure of dehydrated **1a** after the removal of guest water molecules from the host framework by controlled the heating condition and crystal structure of the re-hydration **1b** with the dehydrated **1a** upon exposure to water vapor at RT. Single-crystal-to-single-crystal transformation experiments were performed by single-crystal X-ray diffraction method, which give the structures of dehydrated $[Cd(C_4O_4)(bipy)(H_2O)_2]$ (**1a**) and rehydrated $\{[Cd(C_4O_4)(bipy)(H_2O)_2] \cdot 3H_2O\}_n$ (**1b**), respectively, with very similar cell parameters (Table 3). The structure of the dehydrated **1a** reveals that the 3D supramolecular architecture framework is nearly the same as that of **1** with the difference only on the nonexistence of guest water molecules. The related bond lengths and angles around the Cd(II) ion are similar to that of **1** with little difference (Table 1). Furthermore, when the dehydrated **1a** were exposed to water vapor at RT, structural determination reveals that **1a** re-absorbed water molecules to generate a rehydrated $\{[Cd(C_4O_4)(bipy)(H_2O)_2] \cdot 3H_2O\}_n$ (**1b**), showing a complete reversibility of de-/rehydration processes in the 3D supramolecular network as shown in Scheme 1. The related bond lengths and angles around Cd(II) ion are also in comparable with those of **1** (Table 1). It is important to note that hydrogen bonds existing between the guest water molecules and the host framework found in the rehydrated **1b** are almost the same as those found in **1** (Table 2). This result indicates that hydrogen bonding interaction play a key role on the SCSC transformation mechanism of reversible de-/rehydration and provide a memorial tracing pathway for the guest water during the guest water absorption process.

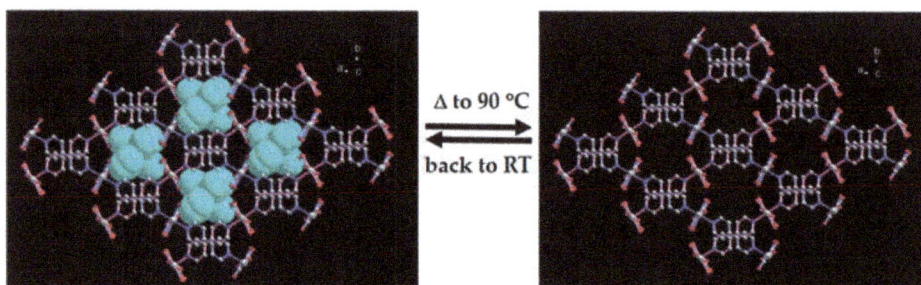

Scheme 1. SCSC structural transformation between **1** and **1a** during the reversible de-/rehydration processes.

Table 3. Crystal data and refinement details of compounds **1**, **1a** and **1b**.

Compound	1	1a	1b
empirical formula	$C_{14}H_{18}Cd_1N_2O_9$	$C_{14}H_{12}Cd_1N_2O_6$	$C_{14}H_{18}Cd_1N_2O_9$
formula mass (g mol^{-1})	470.70	416.66	470.70
crystal system	Monoclinic	Monoclinic	Monoclinic
space group	$C\,2/c$	$C\,2/c$	$C\,2/c$
a (Å)	20.3421(8)	19.7226(10)	20.3589(12)
b (Å)	11.6164(5)	11.7865(5)	11.6195(6)
c (Å)	8.3198(3)	8.1877(3)	8.3196(4)
α (deg)	90	90	90
β (deg)	113.7456(15)	112.1639(15)	113.7461(18)
γ (deg)	90	90	90
V (Å^3)	1799.55(12)	1762.68(13)	1801.47(17)
Z	4	4	4
T (K)	150(2)	340(2)	150(2)
D_{calcd} (g cm^{-3})	1.737	1.570	1.736
μ (mm^{-1})	1.263	1.267	1.262
θ range (deg)	2.066–27.474	2.230–27.484	3.011–27.491
total no. of data collected	5780	5760	6113
no. of unique data	2070	2028	2066
no. of obsd data ($I > 2\sigma(I)$)	1681	1594	1649
R_{int}	0.0334	0.0318	0.0274
refine params	145	114	145
R_1, wR_2 ($I > 2\sigma(I)$) [1]	0.0360, 0.0924	0.0389, 0.0733	0.0250, 0.0482
R_1, wR_2 (all data) [1]	0.0464, 0.1003	0.0551, 0.0800	0.0376, 0.0537
GOF [2]	1.085	1.191	1.095

[1] $R_1 = \Sigma||F_o - F_c||/\Sigma|F_o|$; $wR_2(F^2) = [\Sigma w|F_o^2 - F_c^2|^2/\Sigma w(F_o^4)]^{1/2}$. [2] GOF $= \{\Sigma[w|F_o^2 - F_c^2|^2]/(n - p)\}^{1/2}$.

2.4. Adsorption Properties of (1)

Encouraged by the structural flexibility of **1**, the gas uptake capacities of the dehydrated (activated) framework are determined. Before the measurements, powder samples of compound **1** were evacuated at 90 °C for 24 h to obtain the activated form **1a** (removal of three guest water molecules). The isotherm obtained with N_2 gas at 78 K revealed a typical Type-II adsorption profile with very low uptake, suggesting only surface adsorption. (Figure 4a). Surprisingly, even though **1** is nonporous towards to N_2, we found that it is porous towards CO_2 at 198 K. The CO_2 adsorption isotherms at 198 K for **1a** exhibit a typical Type-I adsorption behavior (Figure 4b) with the adsorption uptake of 18.3 cm^3 g^{-1}.

The adsorption capability of **1** for CO_2 gases is low, but significant. In order to understand the interactions of guest water molecules with the host framework, the water vapor adsorption behavior of the activated **1a** under ambient condition is measured. The activated **1a** show high H_2O uptake (~43.5 cm^3 g^{-1}) at lower P/P_0 (0.0–0.1), and the final uptake amount reaches 82.6 cm^3 g^{-1}, which corresponds to 2.6 molecules of H_2O per formula unit (Figure 4c).

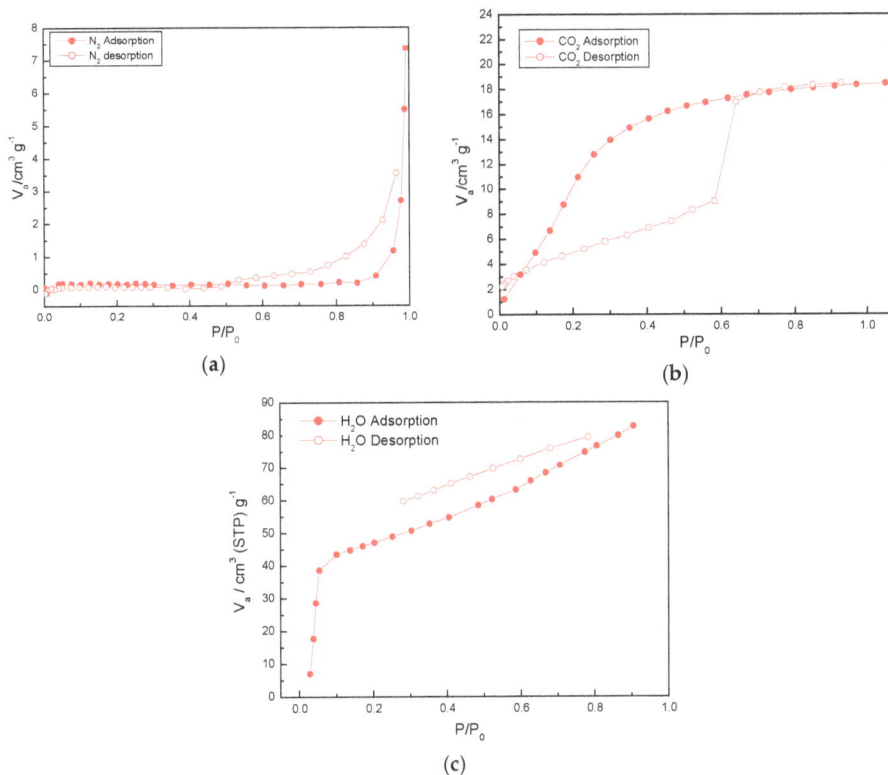

Figure 4. (**a**) N_2 adsorption and desorption isotherms measured at 77 K; (**b**) CO_2 adsorption and desorption isotherms measured at 198 K; and (**c**) H_2O adsorption and desorption isotherms measured at 298 K.

3. Materials and Methods

3.1. Materials and Physical Techniques

All chemicals were of reagent grade and were used as commercially obtained without further purification. Elementary analyses (carbon, hydrogen, and nitrogen) were performed using a Perkin-Elmer 2400 elemental analyzer. IR spectra were recorded on a Nicolet Fourier Transform IR, MAGNA-IR 500 spectrometer in the range of 500–4000 cm^{-1} using the KBr disc technique. Thermogravimetric analysis (TGA) of compounds **1** was performed on a computer-controlled Perkin-Elmer 7 Series/UNIX TGA7 analyzer. Single-phased powder samples were loaded into alumina pans and heated with a ramp rate of 5 °C/min from room temperature to 800 °C under a nitrogen atmosphere. The adsorption isotherm of N_2 (77 K) and CO_2 (200 K) for **1** was measured in the gaseous state by using BELSORP-max volumetric adsorption equipment from BEL, Osaka, Japan. In the sample cell (~1.8 cm^3) maintained at $T \pm 0.03$ K was placed the adsorbent sample (~100–150 mg), which has been prepared at 90 and 180 °C for **1** and 10^{-2} Pa for about 24 h prior to measurement of the isotherm. The adsorbate was placed into the sample cell, and then the change of pressure was monitored and the degree of adsorption was determined by the decrease of pressure at equilibrium state. All operations were through automatically computer-controlled.

3.2. Synthesis of {[Cd(C₄O₄)(bipy)(H₂O)₂]·3H₂O}∞ (1), Dehydrated {[Cd(C₄O₄)(bipy)(H₂O)₂]}∞ (1a) and Rehydrated {[Cd(C₄O₄)(bipy)(H₂O)₂]·3H₂O}∞ (1b)

An ethanol/H_2O solution (1:1, 50 mL) of $H_2C_4O_4$ (0.1419 g, 0.0012 mol) was added to an ethanol/water (1:1, 100 mL) solution of $Cd(NO_3)_2 \cdot 4H_2O$ (0.4107 g, 0.0025 mol) and 4,4′-bipyridine (0.3904 g, 0.0013 mmol) at room temperature. After standing for one week, colorless needle-like crystals of **1** (yield, 0.412g 70.4%) were obtained which are suitable for X-ray diffraction analysis. Anal. Calc. for $C_{14}H_{12}N_2O_6Cd_1$ (**1**): C 35.21, N 5.74, H 3.95; Found: C 35.69, N 5.94, H 3.82. IR (KBr pellet): ν = 1603 (m), 1471 (s), 1413 (s), 1323 (m), 1221 (m), 1067 (m), 1006 (m), 672 (s), 628 (vs) cm^{-1}. The fresh crystals of **1** are heated at 90 °C to obtain dehydrated crystals {[Cd(C₄O₄)(bipy)(H₂O)₂]}∞ (**1a**). The dehydrated crystals **1a** are then stand at RT for one day to obtain rehydrated crystals {[Cd(C₄O₄)(bipy)(H₂O)₂]·3H₂O}∞ (**1b**).

3.3. Crystallographic Data Collection and Refinements

Single-crystal structure analysis for compound **1**, **1a**, and **1b** were performed out on a Siemens SMART diffractomerter (Taipei, Taiwan) with a CCD detector with Mo radiation (λ = 0.71073 Å) at 150 K for **1** and **1b** and at 340 K for **1a**, respectively. A preliminary orientation matrix and unit cell parameters were determined from three runs of 15 frames each, each frame correspond to a 0.3° scan in 10 s, following by spot integration and least-squares refinement. For each structure, data were measured using ω scans of 0.3° per frame for 20 s until a complete hemisphere had been collected. Cell parameters were retrieved using SMART [56] software (Bruker SAINT) and refined with SAINT (Bruker SAINT) [57] on all observed reflections. Data reduction was performed with the SAINT [58] software package and corrected for Lorentz and polarization effects. Absorption corrections were applied with the program SADABS (Bruker, 2016) [58]. Direct phase determination and subsequent difference Fourier map synthesis yielded the positions of all atoms, which were subjected to anisotropic refinements for non-hydrogen atoms and isotropic for hydrogen atoms. The final full-matrix, least-squares refinement on F^2 was applied for all observed reflections [$I > 2\sigma(I)$]. All calculations were performed by using the SHELXL-2014/7 software package [59]. Crystal data and details of the data collection and structure refinements for **1–4** are summarized in Table 1. CCDC-1584569, 1584570, and 1584571 for **1**, **1a**, and **1b** contains the supplementary crystallographic data for this paper. These data can be obtained free of charge at www.ccdc.cam.ac.uk/conts/retrieving.html (or from the Cambridge Crystallographic Data Centre, 12, Union Road, Cambridge CB2 1EZ, UK; fax: (internat.) +44-1223/336-033; email: deposit@ccdc.cam.ac.uk).

3.4. In Situ X-ray Powder Diffraction

Variable temperature synchrotron powder X-ray diffraction data were collected at the Taiwan Photon Source of the National Synchrotron Radiation Research Center (TPS 09A, Hsinchu, Taiwan). The 15 keV X-ray source is delivered from an in-vacuum undulator (IU22, Hsinchu, Taiwan) and the powder diffraction patterns were recorded by a position-sensitive detector, MYTHEN 24K (Hsinchu, Taiwan), covering a 2θ range of 120°. All powder samples were loaded into 0.3 mm capillary. To obtain better random orientation, the capillary was rotated at 400 RPM during data collection. For high temperature experiment, a hot air gas blower was placed 2 mm under the sample. Due to the small gaps between detector modules, the two data sets were collected 2° apart with 60 s exposure time and the data were merged and gridded to give a continuous dataset.

3.5. Measurements of Gas Adsorption of Activated **1a**

The adsorption isotherms of N_2 (77 K), CO_2 (198 K) and H_2O (298 K) were measured in the gaseous state by using BELSORP-max volumetric adsorption equipment from BEL, Osaka, Japan. In the sample cell (~1.8 cm^3) maintained at T ± 0.03 K was placed the adsorbent sample (~100–150 mg), which has been prepared at 90 °C for **1a** and 10^{-2} Pa for about 24 h prior to measurement of the isotherm. The adsorbate was placed into the sample cell, and then the change of pressure was monitored and the degree of adsorption was determined by the decrease of pressure at equilibrium state. All operations were through automatically computer-controlled.

4. Conclusions

A 3D supramolecular network, $\{[Cd(C_4O_4)(bipy)(H_2O)_2] \cdot 3H_2O\}_n$ (**1**), constructed by 2D-layered MOFs has been successfully synthesized. Adjacent 2D layers were then self-assembled via the mutually parallel and interpenetrating manners to form a 3D supramolecular network. Hydrogen bonding interaction among uncoordinated oxygen of squarate, coordinated water molecules in the host framework and the guest water molecules provide extra-energy on the stabilization of the 3D supramolecular architecture of **1**. The guest water molecules in the as-synthesized samples can be removed readily by heating the sample to obtain dehydrated $[Cd(C_4O_4)(bipy)(H_2O)_2]_n$ (**1a**), and this resulting microporous material is highly robust and chemically inert. Noteworthy, compound **1** shows an interesting SCSC structural transformation between the dehydrated $[Cd(C_4O_4)(bipy)(H_2O)_2]$ **1a** and rehydrated $\{[Cd(C_4O_4)(bipy)(H_2O)_2] \cdot 3H_2O\}_n$ **1b** during the reversible thermal re-/dehydration processes. It is also worthy to note that the dehydrated **1a** selectively adsorbs CO_2 over N_2. Moreover, the high water adsorption uptake of the dehydrated **1a** is attributed to the strong hydrogen-bonding affinity of water molecules with the uncoordinated O atoms of the squarate ligands in the host framework.

Acknowledgments: The authors wish to thank the Ministry of Science and Technology and Soochow University, Taiwan for financial support.

Author Contributions: Chih-Chieh Wang conceived and designed the experiments; Szu-Yu Ke, Kuan-Ting Chen, Tzu-Heng Wang, and Yi-Fang Hsieh performed the experiments, including synthesis, structural characterization, EA, IR, TG analysis, and gas-absorption measurements of compounds; Gene-Hsiang Lee contributed to the single-crystal X-ray data collection and structural analysis of compounds 1; Yu-Chun Chuang contributed to the powder X-ray diffraction measurements of compound 1 by synchrotron radiation; and Chih-Chieh Wang and Yu-Chun Chuang wrote the paper.

References

1. Batten, S.R.; Champness, N.R.; Chen, X.M.; Garcia-Martinez, J.; Kitagawa, S.; Öhrström, L.; O'Keeffe, M.; Suh, M.P.; Reedijk, J. Terminology of metal-organic frameworks and coordination polymers. *Pure Appl. Chem.* **2013**, *85*, 1715–1724. [CrossRef]
2. Batten, S.R.; Champness, N.R.; Chen, X.M.; Garcia-Martinez, J.; Kitagawa, S.; Öhrström, L.; O'Keeffe, M.; Suh, M.P.; Reedijk, J. Coordination polymers, metal-organic frameworks and the need for terminology guidelines. *CrystEngComm* **2012**, *14*, 3001–3004. [CrossRef]
3. Li, B.; Chrzanowski, M.; Zhang, Y.; Ma, S. Applications of metal-organic frameworks featuring multi-functional sites. *Coord. Chem. Rev.* **2016**, *307*, 106–129. [CrossRef]
4. Zhang, X.; Wang, W.; Hu, Z.; Wang, G.; Uvdal, K. Coordination polymers for energy transfer: Preparations, properties, sensing applications, and perspectives. *Coord. Chem. Rev.* **2015**, *284*, 206–235. [CrossRef]
5. Bradshaw, D.; Claridge, J.B.; Cussen, E.J.; Prior, T.J.; Rosseinsky, M.J. Design, Chirality, and Flexibility in Nanoporous Molecule-Based Materials. *Acc. Chem. Res.* **2015**, *38*, 273–282. [CrossRef] [PubMed]
6. Silva, P.; Vilela, S.M.F.; Tome, J.P.C.; Paz, F.A.A. Multifunctional metal–organic frameworks: From academia to industrial applications. *Chem. Soc. Rev.* **2015**, *44*, 6774–6803. [CrossRef] [PubMed]

7. Li, S.; Huo, F. Metal-organic framework composites: From fundamentals to applications. *Nanoscale* **2015**, *7*, 7482–7501. [CrossRef] [PubMed]

8. Janiak, C.; Vieth, J.K. MOFs, MILs and more: Concepts, properties and applications for porous coordination networks (PCNs). *New J. Chem.* **2010**, *34*, 2366–2388. [CrossRef]

9. Serre, C.; Millange, F.; Thouvenot, C.; Nogues, M.; Marsolier, G.; Louër, D.; Forey, G. Very Large Breathing Effect in the First Nanoporous Chromium(III)-Based Slids: MIL-53 or CrIII(OH)·{O_2C–C_6H_4–CO_2}·{HO_2C–C_6H_4–CO_2H}$_x$·H_2O_y. *J. Am. Chem. Soc.* **2002**, *124*, 13519–13526. [CrossRef] [PubMed]

10. Kitaura, R.; Seki, K.; Akiyama, G.; Kitagawa, S. Porous coordination-polymer crystals with gated channels specific for supercritical gases. *Angew. Chem. Int. Ed.* **2003**, *42*, 428–431. [CrossRef] [PubMed]

11. Loiseau, T.; Serre, C.; Huguenard, C.; Fink, G.; Taulelle, F.; Henry, M.; Bataille, T.; Forey, G. A rationale for the large breathing of the porous aluminum terephthalate (MIL-53) upon hydration. *Chem. Eur. J.* **2004**, *10*, 1373–1382. [CrossRef] [PubMed]

12. Yamada, K.; Yagishita, S.; Tanaka, H.; Tohyama, K.; Adachi, K.; Kaizaki, S.; Kumagai, H.; Inoue, K.; Kitaura, R.; Chang, H.C.; et al. Metal-Complex Assemblies Constructed from the Flexible Hinge-Like Ligand H2bhnq: Structural Versatility and Dynamic Behavior in the Solid State. *Chem. Eur. J.* **2004**, *10*, 2647–2660. [CrossRef] [PubMed]

13. Takamizawa, S.; Nakata, E.; Saito, T.; Kojima, K. Structural determination of physisorbed sites for CO_2 and Ar gases inside an organometallic framework. *CrystEngComm* **2003**, *5*, 411–413. [CrossRef]

14. Takamizawa, S.; Nakata, E.; Yokoyama, H.; Mochizuki, K.; Mori, W. Carbon dioxide inclusion phases of a transformable 1D coordination polymer host [$Rh_2(O_2CPh)_4(pyz)$]$_n$. *Angew. Chem. Int. Ed.* **2003**, *42*, 4331–4334. [CrossRef] [PubMed]

15. Takamizawa, S.; Nakata, E.; Saito, T. Gas inclusion crystal between 1-D coordination polymer and nitrous oxide: Occurrence of crystal phase transition induced by a slight amount of fluid guests inside host. *Inorg. Chem. Commun.* **2003**, *6*, 1415–1418. [CrossRef]

16. Takamizawa, S.; Nakata, E.; Saito, T. Single-crystal adsorbents: A new observation field for light aggregates. *Angew. Chem. Int. Ed.* **2004**, *43*, 1368–1371. [CrossRef] [PubMed]

17. Takamizawa, S.; Nakata, E.; Saito, T. Structural determination of copper(II) benzoate–pyrazine containing carbon dioxide molecules. *Inorg. Chem. Commun.* **2004**, *7*, 1–3. [CrossRef]

18. Takamizawa, S.; Nakata, E.; Saito, T. Saito, Inclusion Formation between 1D Coordination Polymer Host and CS2 through Vapor Adsorption. *Chem. Lett.* **2004**, *33*, 538–539. [CrossRef]

19. Takamizawa, S.; Nakata, E. Direct observation of H_2 adsorbed state within a porous crystal by single crystal X-ray diffraction analysis. *CrystEngComm* **2005**, *7*, 476–479. [CrossRef]

20. Takamizawa, S.; Nakata, E.; Saito, T.; Akatsuka, T. Generation of a Plastic Crystal Including Methane Rotator within Metal-Organic Cavity by Forcible Gas Adsorption. *Inorg. Chem.* **2005**, *44*, 1362–1366. [CrossRef] [PubMed]

21. Takamizawa, S.; Nakata, E.; Akatsuka, T. Magnetic Behavior of a 1D Molecular-Oxygen System Included within a Transformable Single-Crystal Adsorbent. *Angew. Chem. Int. Ed.* **2006**, *45*, 2216–2221. [CrossRef] [PubMed]

22. Kawano, M.; Kobayashi, Y.; Ozeki, T.; Fujita, M. Direct Crystallographic Observation of a Coordinatively Unsaturated Transition-Metal Complex in situ Generated within a Self-Assembled Cage. *J. Am. Chem. Soc.* **2006**, *128*, 6558–6559. [CrossRef] [PubMed]

23. Yoshizawa, M.; Tamura, M.; Fujita, M. Diels-alder in aqueous molecular hosts: Unusual regioselectivity and efficient catalysis. *Science* **2006**, *312*, 251–254. [CrossRef] [PubMed]

24. Kachi-Terajima, C.; Akatsuka, T.; Kohbara, M.; Takamizawa, S. Structural and Magnetic Study of N_2, NO, NO_2, and SO_2 Adsorbed within a Flexible Single-Crystal Adsorbent of [$Rh_2(bza)_4(pyz)$]$_n$. *Chem. Asian J.* **2007**, *2*, 40–50. [CrossRef] [PubMed]

25. Takamizawa, S.; Kachi-Terajima, C.; Akatsuka, T.; Kohbara, M.; Jin, T. Alcohol-Vapor Inclusion in Single-Crystal Adsorbents [$MII2(bza)4(pyz)$]n (M = Rh, Cu): Structural Study and Application to Separation Membranes. *Chem. Asian J.* **2007**, *2*, 837–848. [CrossRef] [PubMed]

26. Takamizawa, S.; Akatsuka, T.; Ueda, T. Gas-Conforming Transformability of an Ionic Single-Crystal Host Consisting of Discrete Charged Components. *Angew. Chem. Int. Ed.* **2008**, *47*, 1689–1692. [CrossRef] [PubMed]

27. Takamizawa, S.; Kohbara, M.; Akatsuka, T.; Miyake, R. Gas-adsorbing ability of tris-ethylenediamine metal complexes (M = Co(III), Cr(III), Rh(III), Ir(III)) as transformable ionic single crystal hosts. *New J. Chem.* **2008**, *32*, 1782–1787. [CrossRef]

28. Pan, Q.H.; Li, J.Y.; Chen, Q.; Han, Y.D.; Chang, Z.; Song, W.C.; Bu, X.H. $[Co(en)_3]_{1/3}[In(ox)_2]\cdot3.5H_2O$: A zeolitic metal-organic framework templated by $Co(en)_3Cl_3$. *Microporous Mesoporous Mater.* **2010**, *132*, 453–457. [CrossRef]

29. Pan, Q.H.; Chen, Q.; Song, W.C.; Hu, T.L.; Bu, X.H. Template-directed synthesis of three new open-framework metal(II) oxalates using Co(III) complex as template. *CrystEngComm* **2010**, *12*, 4198–4204. [CrossRef]

30. Bertani, R.; Sgarbossa, P.; Venzo, A.; Lelj, F.; Amati, M.; Resnati, G.; Pilati, T.; Metrangolo, P.; Terraneo, G. Halogen bonding in metal-organic-supramolecular networks. *Coord. Chem. Rev.* **2010**, *254*, 677–695. [CrossRef]

31. Ingleson, M.J.; Bacsa, J.; Rosseninsky, M.J. Homochiral H-bonded proline based metal organic frameworks. *Chem. Commun.* **2007**, 3036–3038. [CrossRef] [PubMed]

32. Wang, C.C.; Yang, C.C.; Yeh, C.T.; Ceng, K.Y.; Chang, P.C.; Ho, M.L.; Lee, G.H.; Shih, W.J.; Sheu, H.S. Reversible Solid-State Structural Transformation of a 1D–2D Coordination Polymer by Thermal De/Rehydration Processes. *Inorg. Chem.* **2011**, *50*, 597–603. [CrossRef] [PubMed]

33. Desiraju, G.R. C–H⋯O and other weak hydrogen bonds. From crystal engineering to virtual screening. *Chem. Commun.* **2005**, 2995–3001. [CrossRef] [PubMed]

34. García-Báez, E.V.; Martínez-Martínez, F.J.; Höpfl, H.; Padilla-Martínez, I.I. π-Stacking Interactions and CH⋯X (X = O, Aryl) Hydrogen Bonding as Directing Features of the Supramolecular Self-Association in 3-Carboxy and 3-Amido Coumarin Derivatives. *Cryst. Growth Des.* **2003**, *3*, 35–45. [CrossRef]

35. Jabiak, C. A critical account on π–π stacking in metal complexes with aromatic nitrogen-containing ligands. *Dalton Trans.* **2000**, 3885–3896. [CrossRef]

36. Claessens, C.G.; Stoddart, J.F. π–π Interaction in self-assembly. *J. Phys. Org. Chem.* **1997**, *10*, 254–272. [CrossRef]

37. Guo, H.; Guo, X.; Batten, S.R.; Song, J.; Song, S.; Dang, S.; Zhang, G.; Tang, J.; Zhang, H. Hydrothermal Synthesis, Structures, and Luminescent Properties of Seven d^{10} Metal-Organic Frameworks Based on 9,9-Dipropylfluorene-2,7-Dicarboxylic Acid (H_2DFDA). *Cryst. Growth Des.* **2009**, *9*, 1394–1401. [CrossRef]

38. Reger, D.L.; Horger, J.J.; Smith, M.D.; Long, G.J.; Grandjean, F. Homochiral, Helical Supramolecular Metal-Organic Frameworks Organized by Strong π⋯π Stacking Interactions: Single-Crystal to Single-Crystal Transformations in Closely Packed Solids. *Inorg. Chem.* **2011**, *50*, 686–704. [CrossRef] [PubMed]

39. Wang, C.C.; Yang, C.C.; Chung, W.C.; Lee, G.H.; Ho, M.L.; Yu, Y.C.; Chung, M.W.; Sheu, H.S.; Shih, C.H.; Cheng, K.Y.; et al. A New Coordination Polymer Exhibiting Unique 2D Hydrogen-Bonded ($H_2O)_{16}$ Ring Formation and Water-Dependent Luminescence Properties. *Chem. Eur. J.* **2011**, *17*, 9232–9241. [CrossRef] [PubMed]

40. Kawamura, A.; Greenwood, A.R.; Filatov, A.S.; Gallagher, A.T.; Galli, G.; Anderson, J.S. Incorporation of Pyrazine and Bipyridine Linkers with High-Spin Fe(II) and Co(II) in a Metal-Organic Framework. *Inorg. Chem.* **2017**, *56*, 3349–3356. [CrossRef] [PubMed]

41. Kondo, A.; Satomi, T.; Azuma, K.; Takeda, R.; Maeda, K. New layered copper 1,3,5-benzenetriphosphonates pillared with N-donor ligands: Their synthesis, crystal structures, and adsorption properties. *Dalton Trans.* **2015**, *44*, 12717–12725. [CrossRef] [PubMed]

42. Botezat, O.; van Leusen, J.; Kravtsov, V.C.; Filippova, I.G.; Hauser, J.; Speldrich, M.; Hermann, R.P.; Kramer, K.W.; Liu, S.X.; Decurtins, S. Interpenetrated (8,3)-c and (10,3)-b Metal-Organic Frameworks Based on {Fe-3(III)} and ((Fe2CoII)-Co-III} Pivalate Spin Clusters. *Cryst. Growth Des.* **2014**, *14*, 4721–4728. [CrossRef]

43. Wang, F.; Jing, X.M.; Zheng, B.; Li, G.H.; Zeng, G.; Huo, Q.S.; Liu, Y.L. Four Cd-Based Metal-Organic Frameworks with Structural Varieties Derived from the Replacement of Organic Linkers. *Cryst. Growth Des.* **2013**, *13*, 3522–3527. [CrossRef]

44. Thuery, P. Sulfonate Complexes of Actinide Ions: Structural Diversity in Uranyl Complexes with 2-Sulfobenzoate. *Inorg. Chem.* **2013**, *52*, 435–447. [CrossRef] [PubMed]

45. Gao, C.Y.; Liu, S.X.; Xie, L.H.; Sun, C.Y.; Cao, J.F.; Ren, Y.H.; Feng, D.; Su, Z.M. Rational design microporous pillared-layer frameworks: Syntheses, structures and gas sorption properties. *CrystEngComm* **2009**, *11*, 177–182. [CrossRef]

46. Mao, H.Y.; Zhang, C.H.; Li, G.; Zhang, H.Y.; Hou, H.W.; Li, L.K.; Wu, Q.G.; Zhu, Y.; Wang, E.B. New types of the flexible self-assembled Metal-Organic coordination polymers constructed by aliphatic dicarboxylates and rigid bidentate nitrogen ligands. *Dalton Trans.* **2004**, *22*, 3918–3925. [CrossRef] [PubMed]

47. Batten, S.R.; Murray, K.S. Structure and magnetism of coordination polymers containing dicyanamide and tricyanomethanide. *Coord. Chem. Rev.* **2003**, *246*, 103–130. [CrossRef]

48. Tong, M.L.; Ye, B.H.; Cai, J.W.; Chen, X.M.; Ng, S.W. Clathration of Two-Dimensional Coordination Polymers: Synthesis and Structures of $[M(4,4'-bpy)_2(H_2O)_2](ClO_4)_2 \cdot (2,4'-bpy)_2 \cdot H_2O$ and $[Cu(4,4'-bpy)_2(H_2O)_2]$ $(ClO_4)_4 \cdot (4,4'-H_2Bpy)$ (M = CdII, ZnII and bpy = Bipyridine). *Inorg. Chem.* **1998**, *37*, 2645–2650. [CrossRef] [PubMed]

49. Näther, C.; Greve, J.; Jeβ, I. New Coordination Polymer Changing Its Color upon Reversible Deintercalation and Reintercalation of Water: Synthesis, Structure, and Properties of Poly[Diaqua-(μ_2-Squarato-O,O')-(μ_{2-4},4'-Bipyridine-N,N')- Manganese(II)] Trihydrate. *Chem. Mater.* **2002**, *14*, 4536–4542. [CrossRef]

50. Greve, J.; Jeβ, I.; Näther, C.J. Synthesis, crystal structures and investigations on the dehydration reaction of the new coordination polymers poly[diaqua-(μ_2-squarato-O,O')-(μ_{2-4},4'-bipyridine-N,N')Me(II)] hydrate (Me = Co, Ni, Fe). *Solid State Chem.* **2003**, *175*, 328–340. [CrossRef]

51. Konar, S.; Corbella, M.; Zangrando, E.; Ribas, J.N.; Chaudhuri, R. The first unequivocally ferromagnetically coupled squarato complex: Origin of the ferromagnetism in an interlocked 3D Fe(II) system. *Chem. Commun.* **2003**, 1424–1425. [CrossRef]

52. Wang, C.C.; Yang, C.H.; Tseng, S.M.; Lee, G.H.; Sheu, H.S.; Phyu, K.W. A New Moisture-Sensitive Metal-Coordination Solids, {$[Cd(C_4O_4)(bipy)(H_2O)_2] \cdot 3H_2O\}_\infty$ (bipy = 4,4'-bipyridine). *Inorg. Chim. Acta* **2004**, *357*, 3759–3764.

53. Ito, M.; West, R. New Aromatic Anions. IV. Vibrational Spectra and Force Constants for $C_4O_4^{-2}$ and $C_5O_5^{-2}$. *J. Am. Chem. Soc.* **1963**, *85*, 2580–2584. [CrossRef]

54. Blatov, V.A.; Shevchenko, A.P.; Serezhkin, V.N. TOPOS3.2: A new version of the program package for multipurpose crystal-chemical analysis. *J. Appl. Crystallogr.* **2000**, *33*, 1193. [CrossRef]

55. Blatov, V.A.; Carlucci, L.; Ciani, G.; Proserpio, D.M. Interpenetrating metal–organic and inorganic 3D networks: A computer-aided systematic investigation. Part I. Analysis of the Cambridge structural database. *CrystEngComm* **2004**, *6*, 377–395. [CrossRef]

56. Smart, V. *4.043 Software for CCD Detector System*; Siemens Analytical Instruments Division: Madison, WI, USA, 1995.

57. Saint, V. *4.035 Software for CCD Detector System*; Siemens Analytical Instruments Division: Madison, WI, USA, 1995.

58. Sheldrick, G.M. *Program for the Refinement of Crystal Structures*; University of Göttingen: Göttingen, Germany, 1993.

59. Sheldrick, G.M. *SHELXTL 5.03 (PC-Version), Program Library for Structure Solution and Molecular Graphics*; Siemens Analytical Instruments Division: Madison, WI, USA, 1995.

crystals

MDPI

Article

Synthesis, Crystal Structure, Gas Absorption, and Separation Properties of a Novel Complex Based on Pr and a Three-Connected Ligand

Jie Sun [1,*], Minghui Zhang [2], Aiyun Wang [1] and Ziwei Cai [1]

[1] School of Life Science, Ludong University, Yantai 264025, China; wanay1977@126.com (A.W.); skycaiziwei@163.com (Z.C.)
[2] College of Science, China University of Petroleum (East China), Qingdao 266580, China; zhangmhupchuaxue@163.com
* Correspondence: jiesunld@163.com

Academic Editor: George E. Kostakis
Received: 29 October 2017; Accepted: 7 December 2017; Published: 11 December 2017

Abstract: A novel Pr complex, constructed from a rigid three-connected H_3TMTA and praseodymium(III) ion, has been synthesized in a mixed solvent system and characterized by X-ray single crystal diffraction, infrared spectroscopy, a thermogravimetric analysis, an element analysis, and powder X-ray diffraction, which reveals that complex **1** crystallizes in a three-dimensional porous framework. Moreover, the thermal stabilities and the fluorescent and gas adsorption and separation properties of complex **1** were investigated systematically.

Keywords: rare earth complex; solvothermal conditions; thermal stabilities; fluorescent property; gas uptake

1. Introduction

During the past few decades, a lot of effort has been devoted to the rational design and synthesis of coordination polymers (CPs) in the field of chemical and material science due to their fascinating architectures and topologies together with their potential applications [1–8]. Besides the N-containing ligands, rigid multi-carboxylate ligands are intriguing components owing to their easily predictable and stable resulting framework [9–17]. Among all of the multi-carboxylate ligands, many C_3-symmetric tricarboxylate ligands have been extensively investigated to construct CPs with interesting architectures and properties, including H_3TATB and H_3BTB (TATB denotes 4,4′,4″-s-triazine-2,4,6-triyltribenzoate and BTB denotes benzene-1,3,5-tribenzoate) [18–20]. At the same time, with its three carboxylate groups almost perpendicular to the central benzene ring, a nonplanar ligand H_3TMTA (TMTA denotes 4,4′,4″-(2,4,6-trimethylbenzene-1,3,5-triyl)tribenzoate) has also been applied to build CPs with appealing topologies [21–23].

On the other hand, thousands of CPs based on the transition metal ions have been intensively investigated. Compared with transition metal ions, there exists a kind of rare earth metal ion, which possesses abundant luminescent properties. It should be pointed out that although quite a lot of coordination complexes have been developed using different ligands in the past years, to the best of our knowledge, porous frameworks built from rigid three-tricarboxylate ligands and rare earth ions are still rare.

In the present paper, a novel rare earth complex was constructed from a rigid three-connected H_3TMTA ligand and a praseodymium(III) ion, $(Pr(TMTA)(H_2O)_2)\cdot[DMF\cdot2EtOH\cdot4H_2O]$ [**1**, H_3TMTA = 4,4′,4″-(2,4,6-trimethylbenzene-1,3,5-triyl)tribenzoic acid]. Interestingly, complex **1** shows permanent porosity and a moderate adsorption heat of CO_2 (21.6 kJ·mol^{-1}), which can be used as a platform for the selective adsorption of CO_2/CH_4 (3.56).

2. Experimental

2.1. Materials and Methods

All chemicals were used as commercially received without further purification. The FT-IR spectra were collected from 400 to 4000 cm^{-1} using the KBr pellet method. The elemental analyses (for C, H, or N) were performed on a Perkin-Elmer 240 elemental analyzer ((PerkinElmer, Billerica, MA, USA). The powder X-ray diffraction measurements were performed with a Bruker AXS D8 Advance instrument (Karlsruhe, Germany). The thermogravimetric analysis was recorded on a Mettler Toledo instrument (Mettler Toledo, Zurich, Swiss). The gas uptake was performed on the surface area analyzer ASAP-2020 (Micromeritics, Norcross, GA, USA).

2.2. Synthesis of [Pr(TMTA)(H₂O)₂]·[DMF·2EtOH·4H₂O] (1)

H₃TMTA (2 mg, 0.0045 mmol) and Pr(NO₃)₃·6H₂O (9.2 mg, 0.02 mmol) were dissolved in mixed solvents, DMF:EtOH:H₂O (v:v:v = 1:1:1; 1 mL). The resulting green solution was sealed in a glass tube, heated to 75 °C in 5 h, kept for 40 h, then slowly cooled to 30 °C in 8 h. The green rod crystals were collected, washed with EtOH, and dried in the air (yield: 40%). Elemental analysis calcd (%) for 1: C 49.84, H 5.88, N 1.57; found: C 48.98, H 5.77, N 1.74%. IR (KBr): ν (cm^{-1}) = 3349 (m), 1618 (m), 1554 (s), 1419 (s), 1367 (s), 1273 (w), 1101 (w), 894 (w), 839 (m), 771 (m), 724 (s), 640 (m).

2.3. X-ray Crystallography

The single-crystal structure of the complex 1 was collected by an Agilent Xcalibur Eos Gemini diffractometer (Agilent Technologies, CA, USA) with a (Cu) X-ray Source (Cu-Kα λ = 1.54184 Å). The multi-scan program SADABS was applied to do the absorption corrections [24]. SHELXS-97 and SHELXL-97 were used to solve and refine the final structure of complex 1 by direct methods [25,26]. PLATON was used to add the symmetry of complex 1. [27]. Table 1 contains the crystallographic details of complex 1 and Table 2 collects the selected bond lengths and angles for complex 1.

Table 1. Crystal data for complex **1**.

Empirical Formula	$C_{30}H_{25}O_8Pr$
Formula weight	654.41
Temperature/K	298.15
Crystal system	monoclinic
Space group	$P2_1/n$
a/Å	9.531(3)
b/Å	16.417(5)
c/Å	27.409(8)
$\alpha/°$	90.00
$\beta/°$	93.098(6)
$\gamma/°$	90.00
Volume/Å³	4282(2)
Z	4
ρ_{calc}mg/mm³	1.015
m/mm⁻¹	1.169
F(000)	1312.0
Index ranges	$-10 \leq h \leq 10, 0 \leq k \leq 18, 0 \leq l \leq 30$
Reflections collected	6198
Independent reflections	6198[R(int) = 0.1019]
Data/restraints/parameters	6198/906/354
Goodness-of-fit on F²	1.002
Final R indexes [I >= 2σ (I)]	R_1 = 0.1012, wR_2 = 0.2613
Final R indexes [all data]	R_1 = 0.1277, wR_2 = 0.2752
Largest diff. peak/hole/e Å⁻³	5.28/−1.63

Table 2. Selected bond lengths (Å) and angles (°) for complex **1**.

Pr1-O1	2.390(8)	Pr1-O1w	2.496(8)	Pr1-O2 [1]	2.384(8)
Pr1-O2w	2.488(8)	Pr1-O3 [2]	2.535(8)	Pr1-O4 [2]	2.570(8)
Pr1-O5 [3]	2.445(8)	Pr1-O6 [4]	2.480(8)	Pr1-O6 [3]	2.967(8)
O1-Pr1-O1w	77.9(3)	O1-Pr1-O2w	78.6(3)	O1-Pr1-O3 [1]	76.5(3)
O1-Pr1-O4 [1]	124.3(3)	O1-Pr1-O5 [2]	155.4(3)	O1-Pr1-O6 [2]	138.1(3)

[1] 1 − X, −Y, −Z; [2] −1/2 + X, −1/2 − Y, −1/2 + Z; [3] −1 + X, 1 + Y, +Z; [4] 1 − X, −1 − Y, −Z.

CCDC 1582391 contains the supplementary crystallographic data of complex **1** for this paper. These data could be obtained free of charge via www.ccdc.cam.ac.uk/conts/retrieving.html (or from the CCDC, 12 Union Road, Cambridge CB2 1EZ, UK; fax: +44 1223 336033; E-mail: deposit@ccdc.cam.ac.uk).

3. Results and Discussion

3.1. Crystal Structure of Complex **1**

Complex **1** was obtained in mixed solvents of DMF:EtOH:H_2O by a hydrothermal reaction of H_3TMTA and $Pr(NO_3)_3 \cdot 6H_2O$ at 75 °C. The single-crystal X-ray analysis shows that complex **1** crystalizes in a monoclinic crystal system with a p21/n space group. The asymmetry unit of complex **1** contains a praseodymium ion, a $TMTA^{3-}$ ligand, and two coordinated water molecules. The Pr-O distances are 2.384(8) Å and 2.967(8) Å, and the distances of Pr-Ow are 2.488(8) Å and 2.496(8) Å, respectively. As shown in Figure 1a, the Pr(III) ion in complex **1** adopts a nine-coordinated mode forming a distorted {PrO$_9$} coordination sphere. It is interesting that the carboxylic groups in **1** adopt three different coordination modes: μ_1-η^1-η^1, μ_2-η^1-η^1, and μ_2-η^1-η^2. The carboxylic groups connect with the Pr(III) ion to form a one-dimensional infinite chain, and then the chains are linked by the $TMTA^{3-}$ ligand to construct a three-dimensional framework (Figure 1b).

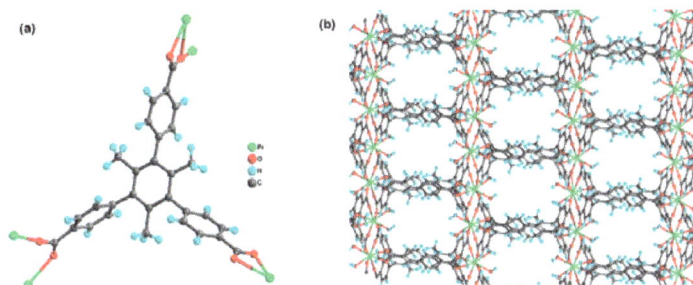

Figure 1. (a) View of the coordination environment around the H_3TMTA ligand and (b) three-dimensional porous framework of **1** viewed along the b axis.

3.2. The Fluorescent Property

Because of the presentation of rare earth ions and a rigid carboxylate group, the luminescent property of complex **1** was tested in the solid state at 298 K. The emission band centered at 362 nm (λ_{ex} = 320 nm) for H_3TMTA, which could be assigned to the electronic transition based on ligand-centered, which means the $\pi^* \rightarrow n$ or $\pi^* \rightarrow \pi$ electronic transition [28]. The emission of complex **1** was observed at 358 nm upon excitation at 320 nm for **1**, which can be attributed to the emission of H_3TMTA ligands (Figure 2). There was no characteristic emission of rare earth ions.

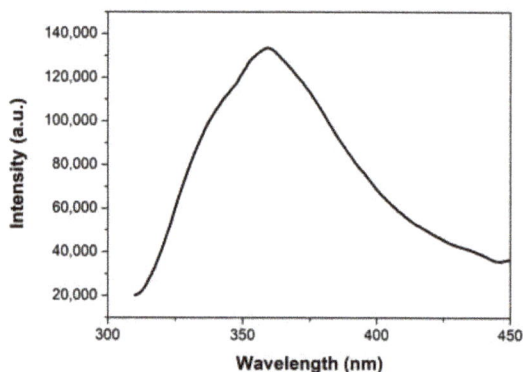

Figure 2. Solid-state fluorescence spectrum of 1 at room temperature.

3.3. Powder X-ray Diffraction Analysis

The powder X-ray diffraction pattern was used to certify the phase purity of complex **1** (Figure 3). Almost all of the peak positions of the simulated and experimental patterns match very well with each other. The preferred orientation of the powder samples accounts for the differences in intensity.

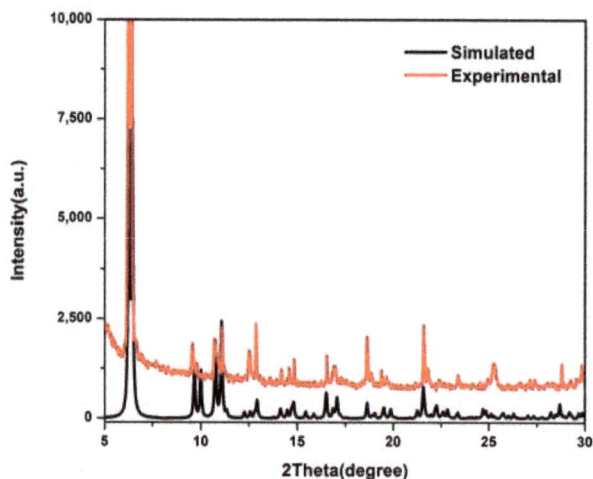

Figure 3. The powder XRD patterns and the simulated pattern from the single-crystal diffraction data for the complex **1**.

3.4. IR Spectra

The FT-IR spectrum of compound **1** was also tested. As depicted in Figure 4, the sharp bands at 1554 cm^{-1} and 1419 cm^{-1} stand for the asymmetric and symmetric stretching vibrations of the carboxylic group, respectively [29].

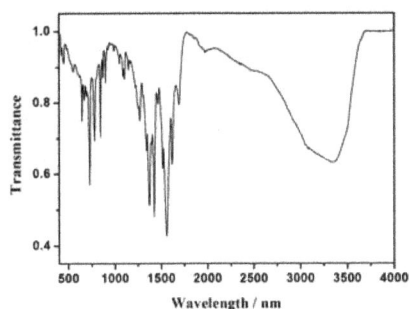

Figure 4. The IR spectra of the complex **1**.

3.5. Thermogravimetric Analyses

As shown in Figure 5, the thermogravimetric analysis (TGA) property of complex **1** was detected under an N_2 atmosphere. Complex **1** has two identifiable weight loss stages: the first stage is similar to the removal of seven uncoordinated and two coordinated solvent molecules (obsd 26.37%, calcd 27.91%), which arises between room temperature and 273 °C. The second stage belongs to the collapse of the framework, which appears at temperatures higher than 500 °C, which means that the present complex **1** shows moderate thermal stability.

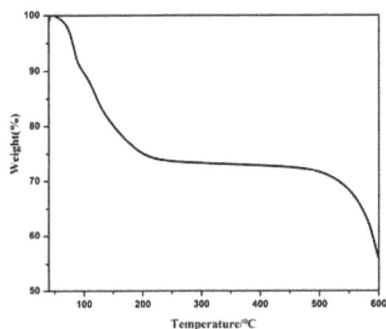

Figure 5. Thermogravimetric analysis (TGA) curves for the complex **1**.

3.6. Gas Sorption and Separation Measurements

Gas adsorption–desorption measurements of N_2, CO_2, CH_4, and H_2 on complex **1** were collected on a Micromeritics ASAP 2020 surface area and pore size analyzer at different temperatures: 77 K (liquid nitrogen bath), 273 K (ice-water bath), and 298 K (room temperature). The Brunauer-Emmett-Teller (BET) surface area and pore size distribution data were calculated from the N_2 adsorption isotherms at 77 K.

The as-synthesized crystals of complex **1** were exchanged three times with dry methanol. The activated phases samples were degassed at 353 K for 10 h for the gas sorption measurements. As can be seen from Figure 6, the active phase is highly crystalline and remains almost identical to its as-synthesized phase. The permanent porosity of complex **1** was confirmed by the reversible N_2 sorption measurements at 77 K and 1 atm, which showed a type I adsorption isotherm performance with a saturated adsorption amount of 106 cm^3 g^{-1}. The values of the Brunauer-Emmett-Teller (BET) and Langmuir surface areas are 327.4 and 422.7 m^2 g^{-1}, respectively, calculated from the N_2 sorption isotherm. The pore size distribution is determined with NLDFT and calculated from N_2 adsorption isotherms at 77 K, corresponding to the pore size of 4.3 Å for complex **1**, which matches well with the crystal data.

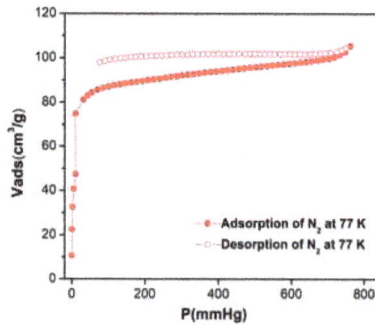

Figure 6. N_2 isotherms at 77 K for complex **1**.

We also tested the low-pressure H_2, CO_2, and CH_4 uptakes of a desolvated sample of complex **1** by using volumetric gas adsorption measurements. Complex **1** can adsorb 89.5 cm^3 g^{-1} of H_2 molecules. Thus, the CO_2 uptake of complex **1** is 26.2 cm$^3 \cdot$g^{-1} (5.158 wt %) at 273 K and 17.6 cm$^3 \cdot$g^{-1} (3.46 wt %) at 298 K under 1 bar, respectively (Figure 7). The adsorption heat (Q_{st}) of CO_2 of complex **1** is 21.6 kJ$\cdot$mol^{-1} calculated from the Clausius-Clapeyron equation, indicating a moderate adsorbate-adsorbant interaction. Furthermore, the CH_4 uptake of complex **1** is 11.6 cm$^3 \cdot$g^{-1} at 273 K and 7.5 cm$^3 \cdot$g^{-1} at 298 K under 1 bar, respectively.

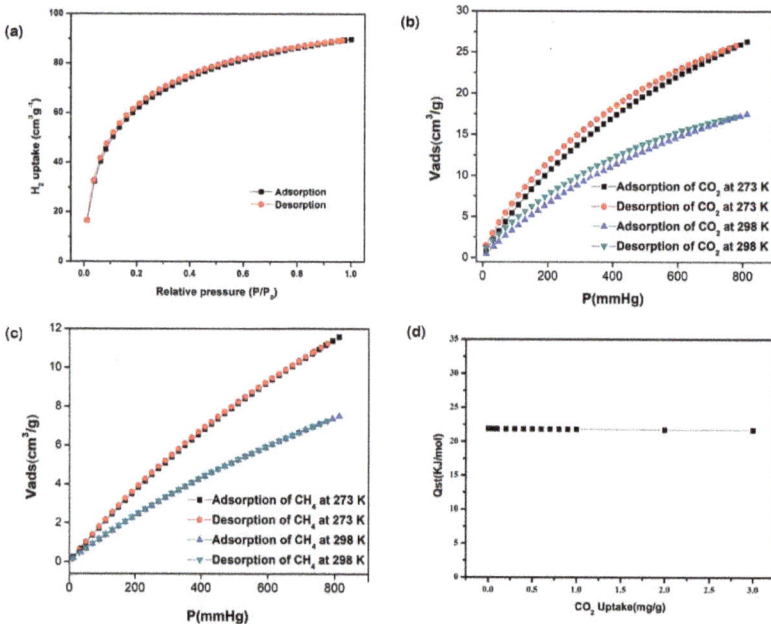

Figure 7. Gas uptakes for complex **1**. (**a**) The H_2 adsorption capacity for complex **1** at 77 K; (**b**) The CO_2 adsorption capacity for complex **1** at 273 and 298 K; (**c**) The CO_2 adsorption capacity for complex **1** at 273 K and 298 K; (**d**) The Q_{st} of complex **1** for CO_2.

Since CO_2 is a dominant component of greenhouse gas and a main contaminant of natural gas, it is meaningful to investigate the capacity of CO_2 and the selectivity of CO_2/CH_4. The higher CO_2 uptake capacity of complex **1** prompted us to further investigate the selectivity of CO_2 adsorption over CH_4. According to the calculation results over a 10:90 and 50:50 CO_2/CH_4 mixed gas, the CO_2/CH_4 selectivitie at 273 K and 298 K are 3.2 and 3.56, respectively. These values are comparable to ZIF-79 (CO_2/CH_4: 5.4) [30], SIFSIX-2-Cu (CO_2/CH_4: 5.3) [31], and PCN-88 (CO_2/CH_4: 5.3) [32] (Figure 8). The results show that compound **1** may be a candidate for CO_2 capture and separation from natural gas.

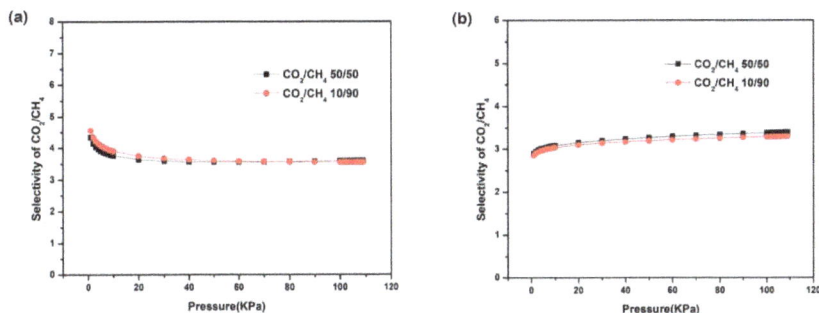

Figure 8. Selective gas adsorption for complex **1**. The CO_2/CH_4 sorption isotherms for complex **1** at 273 K (**a**) and 298 K (**b**) calculated by the IAST method for two CO_2 concentration.

4. Conclusions

In conclusion, A novel Pr complex, constructed from a rigid three-connected H_3TMTA and a praseodymium(III) ion, has been constructed under solvothermal conditions. Thus, the thermal stabilities and the fluorescent and gas adsorption and separation properties of complex **1** were investigated systematically. Complex **1** can be used as a candidate for CO_2 capture and separation from natural gas.

Acknowledgments: We gratefully thank the National Natural Science Foundation of China (No. 21401096) and Open Funds for Key Laboratory of Marine Biotechnology in Colleges and Universities of Shandong Province for the financial support.

Author Contributions: Jie Sun and Minghui Zhang designed the experiments; Aiyun Wang performed the experiments; Ziwei Cai analyzed the data and Jie Sun wrote the paper.

Conflicts of Interest: The authors declare no conflict of interest.

References

1. Forgan, R.S.; Smaldone, R.A.; Gassensmith, J.J.; Furukawa, H.; Cordes, D.B.; Li, Q.; Wilmer, C.E.; Botros, Y.Y.; Snurr, R.Q.; Slawin, A.M.Z.; et al. Nanoporous carbohydrate metal-organic frameworks. *J. Am. Chem. Soc.* **2012**, *134*, 406–417. [CrossRef] [PubMed]
2. Zheng, S.-T.; Bu, J.T.; Li, Y.; Wu, T.; Zuo, F.; Feng, P.; Bu, X. Pore space partition and charge separation in cage-within-cage indium–organic frameworks with high CO_2 uptake. *J. Am. Chem. Soc.* **2010**, *132*, 17062–17064. [CrossRef] [PubMed]
3. Bloch, E.D.; Queen, W.L.; Krishna, R.; Zadrozny, J.M.; Brown, C.M.; Long, J.R. Hydrocarbon separations in a metal-organic framework with open iron(II) coordination sites. *Science* **2012**, *335*, 1606–1610. [CrossRef] [PubMed]
4. Sumida, K.; Rogow, D.L.; Mason, J.A.; McDonald, T.M.; Bloch, E.D.; Herm, Z.R.; Bae, T.-H.; Long, J.R. Carbon dioxide capture in metal-organic frameworks. *Chem. Rev.* **2012**, *112*, 724–781. [CrossRef] [PubMed]
5. Suh, M.P.; Park, H.J.; Prasad, T.K.; Lim, D.W. Hydrogen storage in metal-organic frameworks. *Chem. Rev.* **2012**, *112*, 782–835. [CrossRef] [PubMed]

6. Li, J.R.; Sculley, J.; Zhou, H.-C. Metal-organic frameworks for separations. *Chem. Rev.* **2012**, *112*, 869–932. [CrossRef] [PubMed]

7. Yoon, M.; Srirambalaji, R.; Kim, K. Homochiral metal-organic frameworks for asymmetric heterogeneous catalysis. *Chem. Rev.* **2012**, *112*, 1196–1231. [CrossRef] [PubMed]

8. Ma, L.; Abney, C.; Lin, W. Enantioselective catalysis with homochiral metal–organic frameworks. *Chem. Soc. Rev.* **2009**, *38*, 1248–1256. [CrossRef] [PubMed]

9. Sun, D.; Ma, S.; Ke, Y.; Collins, D.J.; Zhou, H.-C. An interweaving MOF with high hydrogen uptake. *J. Am. Chem. Soc.* **2006**, *128*, 3896–3897. [CrossRef] [PubMed]

10. Ma, S.; Sun, D.; Ambrogio, M.; Fillinger, J.A.; Parkin, S.; Zhou, H.-C. Framework-catenation isomerism in metal–organic frameworks and its impact on hydrogen uptake. *J. Am. Chem. Soc.* **2007**, *129*, 1858–1859. [CrossRef] [PubMed]

11. Ma, S.; Wang, X.-S.; Yuan, D.; Zhou, H.-C. A coordinatively linked Yb metal-organic framework demonstrates high thermal stability and uncommon gas-adsorption selectivity. *Angew. Chem. Int. Ed.* **2008**, *47*, 4130–4133. [CrossRef] [PubMed]

12. Ma, S.; Yuan, D.; Wang, X.-S.; Zhou, H.-C. Microporous lanthanide metal-organic frameworks containing coordinatively linked interpenetration: Syntheses, gas adsorption studies, thermal stability analysis, and photoluminescence investigation. *Inorg. Chem.* **2009**, *48*, 2072–2077. [CrossRef] [PubMed]

13. Ma, S.; Yuan, D.; Chang, J.-S.; Zhou, H.-C. Investigation of gas adsorption performances and H₂ affinities of porous metal-organic frameworks with different entatic metal centers. *Inorg. Chem.* **2009**, *48*, 5398–5402. [CrossRef] [PubMed]

14. Chen, B.; Eddaoudi, M.; Hyde, S.T.; O'Keeffe, M.; Yaghi, O.M. Interwoven metal-organic framework on a periodic minimal surface with extra-large pores. *Science* **2001**, *291*, 1021–1023. [CrossRef] [PubMed]

15. Kim, J.; Chen, B.; Reineke, T.M.; Li, H.; Eddaoudi, M.; Moler, D.B.; O'Keeffe, M.; Yaghi, O.M. Assembly of metal–organic frameworks from large organic and inorganic secondary building units: New examples and simplifying principles for complex structures. *J. Am. Chem. Soc.* **2001**, *123*, 8239–8247. [CrossRef] [PubMed]

16. Chae, H.K.; Siberio-Perez, D.Y.; Kim, J.; Go, Y.; Eddaoudi, M.; Matzger, A.J.; O'Keeffe, M.; Yaghi, O.M. A route to high surface area, porosity and inclusion of large molecules in crystals. *Nature* **2004**, *427*, 523–527. [CrossRef] [PubMed]

17. Gedrich, K.; Senkovska, I.; Klein, N.; Stoeck, U.; Henschel, A.; Lohe, M.R.; Baburin, I.A.; Mueller, U.; Kaskel, S. A highly porous metal-organic framework with open nickel sites. *Angew. Chem. Int. Ed.* **2010**, *49*, 8489–8492. [CrossRef] [PubMed]

18. Yang, X.; Lin, X.; Zhao, Y.; Zhao, Y.; Yan, D. Lanthanide metal-organic framework microrods: Colored optical waveguides and chiral polarized emission. *Angew. Chem. Int. Ed.* **2017**, *56*, 7853–7857. [CrossRef] [PubMed]

19. Cui, Y.; Yue, Y.; Qian, G.; Chen, B. Luminescent functional metal-organic frameworks. *Chem Rev.* **2012**, *112*, 1126–1162. [CrossRef] [PubMed]

20. Feng, X.; Guo, N.; Chen, H.; Wang, H.; Yue, L.; Chen, X.; Ng, S.; Liu, X.; Ma, L.; Wang, L. A series of anionic host coordination polymers based on azoxybenzene carboxylate: Structures, luminescence and magnetic properties. *Dalton Trans.* **2017**, *46*, 14192–14200. [CrossRef] [PubMed]

21. Zhao, X.; He, H.; Dai, F.; Sun, D.; Ke, Y. Supramolecular isomerism in honeycomb metal–organic frameworks driven by CH . . . π interactions: Homochiral crystallization from an achiral ligand through chiral inducement. *Inorg. Chem.* **2010**, *49*, 8650–8652. [CrossRef] [PubMed]

22. Zhao, X.; Dou, J.; Sun, D.; Cui, P.; Sun, D.; Wu, Q. A porous metal-organic framework (MOF) with unusual 2D→3D polycatenation based on honeycomb layers. *Dalton Trans.* **2012**, *41*, 1928–1930. [CrossRef] [PubMed]

23. Zhao, X.; Liu, F.; Zhang, L.; Sun, D.; Wang, R.; Ju, Z.; Yuan, D.; Sun, D. Achieving a rare breathing behavior in a polycatenated 2D to 3D net through a pillar-ligand extension strategy. *Chem. Eur. J.* **2014**, *20*, 649–652. [CrossRef] [PubMed]

24. Bruker. *SMART, SAINT and SADABS*; Bruker AXS Inc.: Madison, WI, USA, 1998.

25. Sheldrick, G.M. *SHELXS-97; Program for X-ray Crystal Structure Determination*; University of Gottingen: Göttingen, Germany, 1997.

26. Sheldrick, G.M. *SHELXL-97; Program for X-ray Crystal Structure Refinement*; University of Gottingen: Göttingen, Germany, 1997.

27. Spek, A.L. *PLATON; A Multipurpose Crystallographic Tool*; Utrecht University: Utrecht, The Netherlands, 2002.

28. Zhang, L.; Guo, J.; Meng, Q.; Wang, R.; Sun, D. Syntheses, structures and characteristics of four metal–organic coordination polymers based on 5-hydroxyisophthalic acid and N-containing auxiliary ligands. *CrystEngComm* **2013**, *15*, 9578–9587. [CrossRef]

29. Nakamoto, K. *Infrared and Raman Spectra of Inorganic and Coordination Compounds*; John Wiley & Sons: New York, NY, USA, 1986.

30. Phan, A.; Doonan, C.J.; Uribe-Romo, F.J.; Knobler, C.B.; O'Keeffe, M.; Yaghi, O.M. Synthesis, structure, and carbon dioxide capture properties of zeolitic imidazolate frameworks. *Acc. Chem. Res.* **2010**, *43*, 58–67. [CrossRef] [PubMed]

31. Nugent, P.; Belmabkhout, Y.; Burd, S.D.; Cairns, A.J.; Luebke, R.; Forrest, K.; Pham, T.; Ma, S.; Space, B.; Wojtas, L.; et al. Porous materials with optimal adsorption thermodynamics and kinetics for CO_2 separation. *Nature* **2013**, *495*, 80–84. [CrossRef] [PubMed]

32. Li, J.R.; Yu, J.; Lu, W.; Sun, L.B.; Sculley, J.; Balbuena, P.B.; Zhou, H.C. Porous materials with pre-designed single-molecule traps for CO_2 selective adsorption. *Nat. Commun.* **2013**, *4*, 1538–1544. [CrossRef] [PubMed]

crystals

MDPI

Article

Solvent Dependent Disorder in M₂(BzOip)₂(H₂O)·Solvate (M = Co or Zn)

Laura J. McCormick [1,2,*] , **Samuel A. Morris** [2,3], **Simon J. Teat** [1], **Alexandra M.Z. Slawin** [2] and **Russell E. Morris** [2]

[1]   Advanced Light Source, Lawrence Berkeley National Laboratory, Berkeley, CA 94720, USA; sjteat@lbl.gov
[2]   School of Chemistry, University of St Andrews, North Haugh, St Andrews, Fife KY16 9ST, UK;
      smorris@ntu.edu.sg (S.A.M.); amzs@st-andrews.ac.uk (A.M.Z.S.); rem1@st-andrews.ac.uk (R.E.M.)
[3]   Facility for Analysis, Characterisation, Testing and Simulation, Nanyang Technological University,
      Singapore 639798, Singapore
*   Correspondence: ljmccormick@lbl.gov; Tel.: +1-510-495-2887

Received: 5 December 2017; Accepted: 19 December 2017; Published: 24 December 2017

Abstract: Coordination polymers derived from 5-benzyloxy isophthalic acid (H₂BzOip) are rare, with only three reported that do not contain additional bridging ligands, of which two M₂(BzOip)₂(H₂O) (M = Co and Zn) are isomorphous. It was hoped that by varying the solvent system in a reaction between H₂BzOip and M(OAc)₂ (M = Co and Zn), from water to a water/alcohol mixture, coordination polymers of different topology could be formed. Instead, two polymorphs of the existing M₂(BzOip)₂(H₂O) (M = Co and Zn) were isolated from aqueous methanol and aqueous ethanol, in which a small number of guest solvent molecules are present in the crystals. These guest water molecules disrupt the hexaphenyl embrace motif, leading to varying degrees of disorder of the benzyl groups.

Keywords: X-ray crystallography; solvothermal; polymorph; disorder

1. Introduction

Coordination polymers containing 5-substituted isophthalates have been, until recently, comparatively rare when compared to those containing the closely related terephthalate [1–15] and trimesate [16–18]. This family of ligands are of interest as they contain two carboxylate coordinating groups in a fixed geometry, while the 5-substituent is remote from these binding groups so that it can act as a structure-directing group without imposing a steric influence on the binding mode of the ligand. We recently reported a series of coordination polymers [19] derived from 5-alkoxy isophthalate ligands, in which the topology of the coordination polymers depended on which short chain alkyl group (ethyl, *n*-propyl, *n*-butyl or *i*-butyl) was present and what solvent was used. Given this solvent dependence of the topology of the coordination frameworks, it was hoped that the same would hold true when an aromatic substituent was used. The 5-benzyloxy isophthalic acid (H₂BzOip, **I**) was selected due to its similarity to the aforementioned 5-alkoxy isophthalic acids. It has been used as a bridging ligand in 11 coordination complexes to date [20–26], although all but three of these [26–28] include additional bridging or terminal polydentate ligands. Two of these, M₂(BzOip)₂(H₂O) (M = Co [26] or Zn [27]) have the same topology as the M₂(ROip)₂(H₂O) (M = Mn, Co or Zn, R = Et, ⁿPr, ⁿBu or ⁱBu) materials reported in 2016, and were both prepared by hydrothermal reaction of H₂BzOip and the corresponding metal nitrate. By varying the solvent system from water to aqueous methanol, ethanol or *iso*propanol, it was found that, although the topology of the resulting M₂(BzOip)₂(H₂O) (M = Co or Zn) crystals did not vary, the level of disorder of the benzyl groups did. Herein, we report the crystal structures of the isostructural compounds M₂(BzOip)₂(H₂O)·solvate (M = Co or Zn) as prepared from aqueous methanol, ethanol and *iso*propanol.

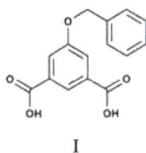

I

2. Results

All compounds were prepared by the solvothermal reaction of H_2BzOip and $M(OAc)_2$ (M = Co or Zn) in a 1:2 mixture of water and alcohol (methanol, ethanol, or *iso*propanol) and formed as purple or colourless needle crystals, respectively. $M_2(BzOip)_2(H_2O)$·solvate (M = Co or Zn) crystallises in the trigonal space group R3 with approximate unit cell dimensions $a \approx 28$ Å and $c \approx 19$ Å. Details of the first crystal structure determination for each compound are presented in Table 1, whilst selected bond lengths and angles for all 18 structural determinations are given in Tables S1 to S36 in the Supporting Materials. The structure determinations of $Zn_2(BzOip)_2(H_2O)$ from *iso*propanol and ethanol are identical to that reported in [27], whilst the structure determination of $Co_2(BzOip)_2(H_2O)$ from *iso*propanol is identical to that reported in [26].

Two crystallographically distinct metal centres and two crystallographically distinct 5-benzyloxy isophthalate ligands are present in the asymmetric unit. One metal centre, M1, adopts a tetrahedral coordination environment in which one oxygen atom from each of the four unique carboxylate groups binds to the metal centre. Each of these carboxylate groups bind in an ($\eta1:\eta2:\mu_2$) fashion, three of which connect M1 and M2 into a discrete unit that is reminiscent of a copper acetate lantern, whilst the fourth connects M1 to an M2 centre in an adjacent lantern-like unit. M2 adopts a square pyramidal geometry comprised of four carboxylate oxygen atoms (each from a distinct carboxylate group), and one water molecule. Both metal centres are also involved in long interactions (see Figure 1) with carboxylate oxygen atoms that are bound to the opposite metal centre that are deemed too long to be a coordination bond (approximate separations M1···O 2.9 and 2.9 Å, and M2···O 3.1 and 3.3 Å for Co and Zn, respectively).

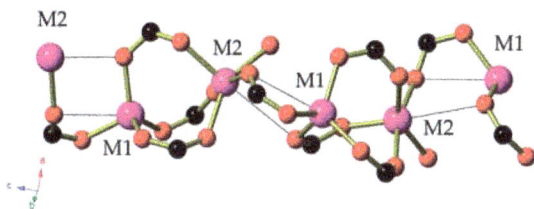

Figure 1. The coordination environment of the metal centres in $Co_2(BzOip)_2(H_2O)$. Long interactions between Co centres and carboxylate oxygen atoms are shown as thin black lines.

The isophthalate moiety of the 5-benzyloxy isophthalate ligand connects the metal centres into a honeycomb framework that may be seen in Figure 2. The panels that constitute the walls of the framework comprise pairs of isophthalate moieties, which are arranged parallel to each other with a plane-to-plane separation of approximately 3.6–3.7 Å. Panels are connected to each other by infinite columns of metal centres and carboxylate groups that extend parallel to the c-axis, leaving large hexagonal channels parallel to the c-axis. The coordinated water molecules form hydrogen bonds to the coordinated carboxylate oxygen atoms and the phenolate oxygen atoms. The benzyloxy groups of the 5-benzyloxy isophthalate ligands project into and completely block these channels, where they form aggregates of six aromatic rings.

Table 1. Crystallographic data for the first run of the various $M_2(BzOip)_2(H_2O)$ crystals.

	Zn from MeOH	Zn from EtOH	Zn from iPrOH	Co from MeOH	Co from EtOH	Co from iPrOH
Molecular Formula	$C_{30.22}H_{23.35}O_{11.45}Zn_2$	$C_{30}H_{22}O_{11}Zn_2$	$C_{30}H_{22}O_{11}Zn_2$	$C_{30.29}H_{23.50}Co_2O_{11.58}$	$C_{30}H_{22.39}Co_2O_{11.19}$	$C_{30}H_{22}Co_2O_{11}$
Temperature	100(2)	173(2)	173(2)	100(2)	100(2)	173(2)
λ (Å)	0.7749	0.71075	0.71075	0.7749	0.7749	0.71075
Crystal System	Trigonal	Trigonal	Trigonal	Trigonal	Trigonal	Trigonal
Space Group	R3	R3	R3	R3	R3	R3
a (Å)	27.827(4)	27.759(3)	27.6973(18)	28.013(3)	27.877(3)	27.797(4)
c (Å)	18.255(2)	18.401(3)	18.2593(18)	18.1437(19)	18.192(2)	18.224(4)
V (Å^3)	12242(4)	12279(3)	12131(2)	12330(3)	12244(3)	12195(4)
Z	18	18	18	18	18	18
ϱ (g cm^{-1})	1.710	1.678	1.698	1.674	1.660	1.658
μ (mm^{-1})	2.309	1.822	1.844	1.609	1.617	1.282
$F(000)$	6413	6300	6300	6334	6227	6192
GooF	1.037	1.059	0.839	1.070	1.056	1.044
Reflections collected/unique/parameters	29,545/8105/467	42,033/6177/394	41,948/6193/394	29,077/8331/471	29,770/8248/441	41,642/6174/396
R_{int}	0.0712	0.0929	0.0862	0.0595	0.0565	0.1225
Final R indices ($I > 2\sigma(I)$)	R1 = 0.0506 wR2 = 0.1476	R1 = 0.0385 wR2 = 0.0708	R1 = 0.0349 wR2 = 0.0575	R1 = 0.0462 wR2 = 0.1373	R1 = 0.0405 wR2 = 0.1109	R1 = 0.0496 wR2 = 0.0960
Final R indices (all data)	R1 = 0.0593 wR2 = 0.1550	R1 = 0.0732 wR2 = 0.0827	R1 = 0.0616 wR2 = 0.0629	R1 = 0.0506 wR2 = 0.1416	R1 = 0.0471 wR2 = 0.1156	R1 = 0.1030 wR2 = 0.1167

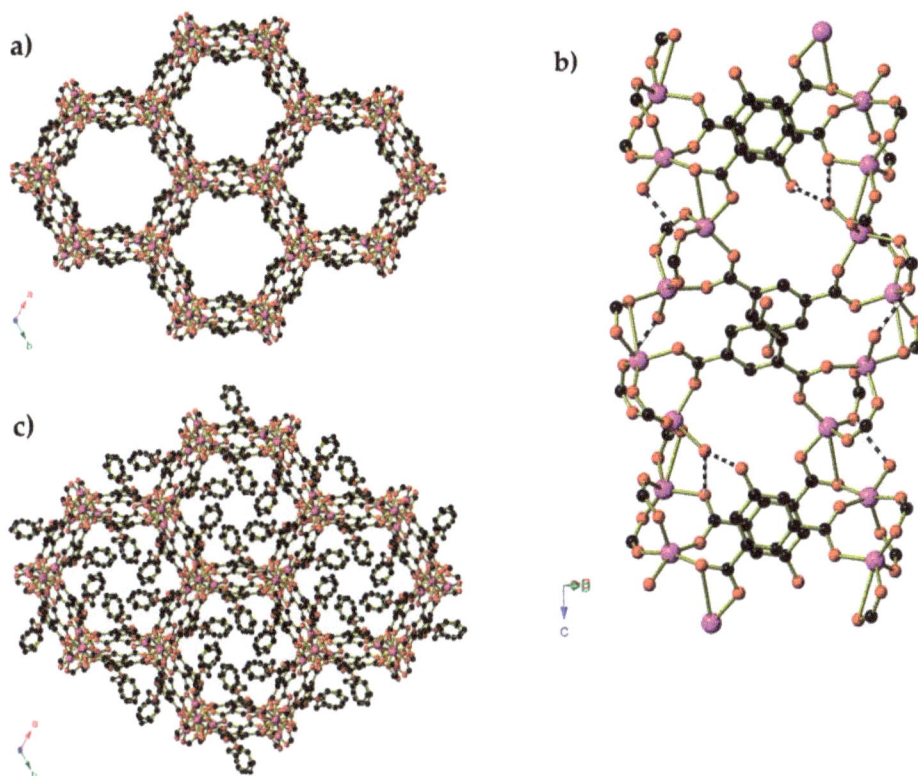

Figure 2. (**a**) A view along the *c*-axis of the honeycomb frameworks of $Co_2(BzOip)_2(H_2O)$ in which the benzyloxy groups have been removed for clarity; (**b**) one segment of the framework wall panels in $Co_2(BzOip)_2(H_2O)$, with the benzyl groups removed for clarity; and (**c**) a view of the full framework of $Co_2(BzOip)_2(H_2O)$.

As there are two crystallographically distinct ligands, two distinct hexaphenyl aggregates are formed (Figure 3). The first is well ordered in all six compounds and the six aromatic rings are involved in a hexaphenyl embrace [29]. Within this aggregate (shown using black carbons atoms and gold bonds in Figure 3), the methylene bridges connecting the benzyl group to the oxo isophthalate core are arranged around the equator of the aggregate. The arrangement of benzyl groups in the second aggregate varies from compound to compound. In both the $Co_2(BzOip)_2(H_2O)$ crystals from ^{i}PrOH, and the $Zn_2(BzOip)_2(H_2O)$ crystal from EtOH and ^{i}PrOH, these benzyl groups are well ordered and participate in a hexaphenyl embrace similar to that seen in the first aggregate, except that the methylene groups are located at the 'poles' of the aggregate. This aggregate is shown in Figure 3a using hatched black carbon atoms. In both the Zn and Co crystals from MeOH and the Co crystal from EtOH, this benzyl group is disordered over multiple orientations to accommodate the presence of partial occupancy solvent molecules. These solvent molecules participate in hydrogen bonding interactions with the carboxylate oxygen atoms and, in some instances, the coordinated water molecules.

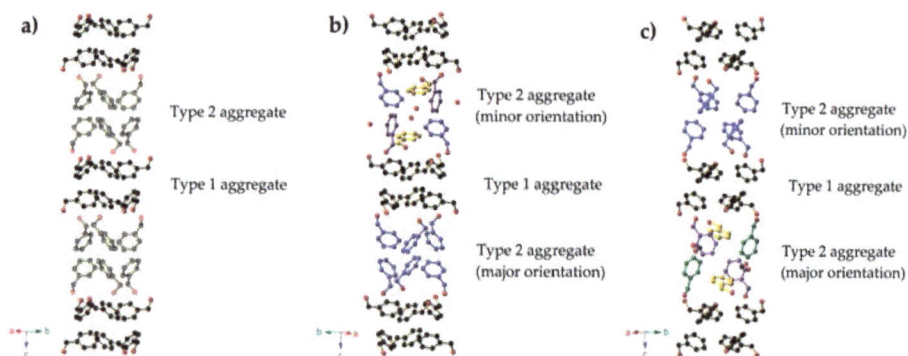

Figure 3. Views of the interactions between benzyl groups that occur in (**a**) $Co_2(BzOip)_2(H_2O)$ prepared in aqueous *iso*propanol; (**b**) $Co_2(BzOip)_2(H_2O) \cdot 0.19H_2O$ prepared in aqueous ethanol; (**c**) $Co_2(BzOip)_2(H_2O) \cdot 0.29H_2O \cdot 0.29MeOH$ prepared in aqueous methanol. In all cases, the benzyl groups shown in black are well ordered and participate in a hexaphenyl embrace. In (**b**) and (**c**), the benzyl groups shown in blue also participate in a hexaphenyl embrace. The hexaphenyl aggregate shown using hatched atoms in (**a**) is fully ordered in the crystals obtained from *iso*propanol, but disordered in the crystals obtained from methanol or ethanol.

In the $Co_2(BzOip)_2(H_2O) \cdot 0.19H_2O$ crystals from aqueous ethanol, the second aggregate of benzyl groups contains one major (shown in blue in Figure 3b) and three minor orientations. The three minor orientations are related by three-fold rotation in the *ab*-plane, and contain two benzyl groups from the major (blue) orientation, and two of each of the minor orientations of the benzyl group (shown in purple and yellow in Figure 3b). Whilst the major orientation of this aggregate results in a hexaphenyl embrace between the benzyl groups, the minor orientations contain a tetraphenyl embrace [30] between the benzyl groups coloured blue and purple in Figure 3b. Partial occupancy water molecules located near to this aggregate and the major orientation of the benzyl groups are positioned such that they are mutually exclusive, and the presence of the water molecules disrupts the hexaphenyl embrace of the major orientation. The ratio of the major to minor orientations of this aggregate are approximately 80:20. A close-up view of the minor orientation is shown in Figure 4a.

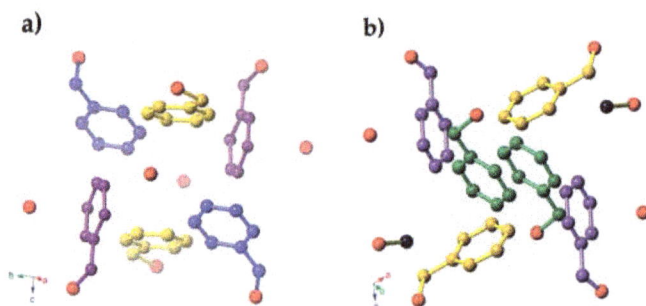

Figure 4. Views of the interactions between benzyl groups that occur in the disordered hexaphenyl aggregates in (**a**) $Co_2(BzOip)_2(H_2O) \cdot 0.19H_2O$ prepared in aqueous ethanol; and (**b**) $Co_2(BzOip)_2(H_2O) \cdot 0.29H_2O\ 0.29MeOH$ prepared in aqueous methanol. Benzyl groups depicted using the same colours are related by symmetry. Hydrogen atoms have been omitted for clarity.

In the $M_2(BzOip)_2(H_2O) \cdot 0.29H_2O \cdot 0.29MeOH$ (M = Co or Zn) crystals, the second aggregate has three 'major' orientations and one minor orientation, and the two have no benzyl groups in common.

The minor orientation is comprised of six symmetry-related benzyl groups arranged into a hexaphenyl embrace (shown in blue in Figure 3c) in a very similar fashion to the major orientation of the second aggregate in $Co_2(BzOip)_2(H_2O)\cdot0.19H_2O$. Conversely, the major orientations of this second orientation are similar to the minor orientation in $Co_2(BzOip)_2(H_2O)\cdot0.19H_2O$, as the three major orientations are related by three-fold symmetry and are comprised of two of each of three further distinct benzyl group sites (shown in purple, green, and yellow in Figures 3c and 4b). The benzyl groups coloured yellow and green in Figure 3c participate in a tetraphenyl embrace. As with $Co_2(BzOip)_2(H_2O)\cdot0.19H_2O$, there are solvent molecules present in the area surrounding the second benzyl aggregate that interferes with the hexaphenyl embrace motif. In these instances, there is a disordered water/methanol molecule that is mutually exclusive of the minor orientation of the aggregate and the green site of the major orientations (Figure 4b). In addition to this, the carbon atom of the methanol molecule lies too close to the purple benzyl ring site, resulting in two water molecules and two methanol molecules surrounding each of the major orientation aggregates. The ratio of the major to minor orientations is approximately 88:12 in the Co derivative, but 68:32 in the Zn derivative.

3. Discussion

In order to confirm that the disorder was not an aberration that occurred only in one crystal, single-crystal diffraction data was collected on three different crystals from the first batch of each of $Co_2(BzOip)_2(H_2O)\cdot0.29H_2O\cdot0.29MeOH$, $Co_2(BzOip)_2(H_2O)\cdot0.19H_2O$, and $Co_2(BzOip)_2(H_2O)$, and two different crystals from a second batch of each. In all cases, the level of disorder was reasonably consistent between different crystals and different batches produced using the same synthetic conditions. In $Co_2(BzOip)_2(H_2O)\cdot0.29H_2O\cdot0.29MeOH$, the occupancy of the minor orientation of the disordered benzyl group (i.e., the one that forms hexaphenyl embraces, shown in blue in Figure 3c) varies from 9.6% to 12.6% in batch 1 and 12.1% to 15.9% in batch 2 (this occupancy was 32.3% in $Zn_2(BzOip)_2(H_2O)\cdot0.23H_2O\cdot0.22MeOH$). In $Co_2(BzOip)_2(H_2O)\cdot0.19H_2O$, the occupancy of the major orientation of the disordered benzyl group (i.e., the one that forms hexaphenyl embraces, shown in blue in Figure 3b) varies from 78.4% to 81.0% in batch 1 and 77.9% to 79.9% in batch 2. Repeat structure determinations were not performed on $Zn_2(BzOip)_2(H_2O)\cdot0.23H_2O\cdot0.33MeOH$. Comparison of the calculated and experimental powder diffraction patterns for each of the 6 compounds is presented in Supporting Materials Figures S1–S6.

The progressive exclusion of solvent from the crystals is reflected in the thermogravimetric analysis (TGA) traces of the $Co_2(BzOip)_2(H_2O)\cdot$solvate series (see Supporting Materials Figures S7–S12), with the mass loss step dropping from 5.07% for the crystals obtained from MeOH, to 3.83% from EtOH, to 2.79% from ^iPrOH, the latter corresponding to the approximate loss of the coordinated water molecule (2.66%). The mass losses from the corresponding $Zn_2(BzOip)_2(H_2O)\cdot$solvate series all lie between 3.47 to 4.01% (MeOH and ^iPrOH, respectively). This suggests that the guest methanol is readily lost from the Zn crystals under ambient conditions.

In all crystals where disorder of the benzyl groups are observed, there are solvent (water and/or methanol) molecules present in the region of the hexaphenyl embraces making it sterically difficult for the type 2 benzyl groups to adopt the hexaphenyl embrace motif. This is noteworthy, as this does not occur when water alone is used as the preparative solvent, despite water being present in far greater amounts. We attribute this to the relative polarities of the solvent mixtures—the most and least polar solvent mixtures (water and aqueous *iso*propanol, respectively) see complete separation of the hydrophobic benzyl groups from the hydrophilic carboxylate groups and coordinated water molecules within the crystals in a manner that is similar to micelle formation. In the intermediate polarity solvent mixtures, aqueous methanol and ethanol, the micelle-type separation of the hydrophilic and hydrophobic sections of the ligand is incomplete as the solvent molecules participate in hydrogen bonding interactions to the phenolic oxygen atom, disrupting the formation of half of the hexaphenyl embrace motifs.

A similar pattern regarding the micelle-type separation of the hydrophilic carboxylate groups and hydrophobic alkyl chains was observed in coordination polymers derived from 5-alkoxy isophthalates. The presence of short alkyl chains in both the solvent system (e.g., MeOH) and the 5-alkoxy substituent (e.g., $EtOip^{2-}$) afforded frameworks in which both alkyl chains and coordinated solvent molecules project into the channels. Frameworks isostructural to $M_2(BzOip)_2(H_2O)$ (M = Co or Zn) were obtained when large hydrophobic alkyl chains were present in the ligand (e.g., $^nBuOip^{2-}$), the solvent system (e.g., neat or aqueous ^iPrOH), or both. Like the branched *iso*butyl chains, the benzyl group used in this study is too bulky to form the $M_6(ROip)_5(OH)_2(H_2O)_4 \cdot xH_2O$ framework in which the alkyl chains and coordinated solvent both project into the same channels. It is worth noting that a related framework of composition $Zn(BzNHip)$ ($BzNHip^{2-}$ = 5-benzylamino isophthalate), which also contains hexagonal channels in which the benzyl substituents participate in hexaphenyl embrace interactions, was prepared from water and does not display any disorder of the benzyl substituents [31,32].

4. Materials and Methods

All reagents were obtained from commercial sources and were used without further purification.

4.1. Dimethyl 5-Hydroxy Isophthalate

Prepared using the literature method [33]. 5-Hydroxy isophthalic acid (9.11 g, 0.050 mol) was heated to reflux in a solution of conc. sulfuric acid (2.8 mL) in methanol (150 mL) for 16 h. The solution was cooled to room temperature, neutralised with saturated $NaHCO_3$, and the solvent removed under reduced pressure. The residue was triturated in water (150 mL), and the inhomogeneous mixture extracted with dichloromethane (4 × 500 mL). The organic layers were collected, dried over magnesium sulfate, and the dichloromethane removed under reduced pressure. Yield: 9.21 g, 0.044 mol, 88%.

4.2. Dimethyl 5-Benzyloxy Isophthalate

Prepared using the literature method [34]. Dimethyl 5-hydroxy isophthalate (6.1 g, 0.029 mol) and freshly pulverised dry potassium carbonate (4.65 g, 0.034 mol) were suspended in a solution of acetone (50 mL) and acetonitrile (60 mL). The solution was allowed to reflux for approximately 1 h, and then benzyl chloride (3.65 g, 0.029 mol) was added dropwise. The mixture was allowed to reflux for a further 16 hours before cooling to room temperature and removing the solvent under reduced pressure. The residue was then completely taken up into a mixture of water (150 mL) and ethyl acetate (400 mL). The organic layer was isolated and washed with water (4 × 50 mL) until the aqueous layer reached pH 7. The organic layer was then washed with brine (50 mL), dried over sodium sulfate and the solvent removed under reduced pressure to afford a pale yellow oil that set into colourless crystals (9.75 g) upon cooling. The crystals were recrystallised from toluene/ethanol to give a pure product. Yield: 6.03 g, 0.020 mol, 70%.

4.3. 5-Benzyloxy Isophthalic Acid

Prepared using the literature method [34]. Dimethyl 5-benzyloxy isophthalate (3.95 g, 0.013 mol) was dissolved in a solution of potassium hydroxide (30.06 g, 0.54 mol) in methanol (300 mL) and the solution refluxed for approximately 2.5 h. The cooled solution was neutralised with 2M HCl. The white precipitate was collected by vacuum filtration, washed with water until the filtrate reached pH 7, and dried in a vacuum desiccator for 2 days. Yield: 3.13 g, 0.011 mol, 87%.

4.4. $M_2(BzOip)_2(H_2O) \cdot$ Solvate (M = Co or Zn)

Into a 23 mL Teflon-lined steel autoclave were placed 5-benzyloxy isophthalic acid (0.272 g, 1.00 mmol), and either $Co(OAc)_2 \cdot 4H_2O$ (251 mg, 1.01 mmol) or $Zn(OAc)_2 \cdot 2H_2O$ (0.220 g, 1.00 mmol), and the solids were suspended in a mixture of water (5 mL) and methanol (10 mL). The sealed autoclaves were placed into a 110 °C oven for 3 days. After this time, the autoclaves were cooled to room temperature and the resulting purple (Co) or colourless (Zn) needle crystals were collected by vacuum filtration and washed with methanol. Yields:

$Co_2(BzOip)_2(H_2O)\cdot0.29MeOH\cdot0.29H_2O$ from methanol: 280 mg, 0.41 mmol, 80%. Elemental analysis Calcd for $C_{30.29}H_{23.74}O_{11.58}Co_2$ C: 52.66 H: 3.46, Found C: 52.53 H: 3.31%.

$Zn_2(BzOip)_2(H_2O)\cdot0.22MeOH\cdot0.23H_2O$ from methanol: 269 mg, 0.38 mmol, 77%. Elemental analysis Calcd for $C_{30.22}H_{23.34}O_{11.45}Zn_2$ C: 51.82 H: 3.36, Found C: 51.60 H: 2.86%.

$Co_2(BzOip)_2(H_2O)\cdot0.19H_2O$ from ethanol: Yields Co: 203 mg, 0.30 mmol, 59%. Elemental analysis Calcd for $C_{30}H_{22.38}O_{11.19}Co_2$ C: 53.01 H: 3.32, Found C: 52.80 H: 3.20%.

$Zn_2(BzOip)_2(H_2O)$ from ethanol: 266 mg, 0.39 mmol, 77%. Elemental analysis Calcd for $C_{30}H_{22}O_{11}Zn_2$ C: 52.28 H: 3.22, Found C: 52.15 H: 3.00%.

$Co_2(BzOip)_2(H_2O)$ from *iso*propanol: 246 mg, 0.36 mmol, 72%. Elemental analysis Calc for $C_{30}H_{22}O_{11}Co_2$ C: 53.27 H: 3.28, Found C: 53.01 3.35%.

$Zn_2(BzOip)_2(H_2O)$ from *iso*propanol: 320 mg, 0.46 mmol, 93%. Elemental analysis Calcd for $C_{30}H_{22}O_{11}Zn_2$ C: 52.28 H: 3.22, Found C: 52.28 H: 3.04%.

4.5. Crystallographic Analyses

All crystals were coated in a protective oil before being transferred to either a Bruker D8 diffractometer (Bruker, Madison, WI, USA) on station 11.3.1 at the Advanced Light Source ($Co_2(BzOip)_2(H_2O)\cdot0.29MeOH\cdot0.29H_2O$ from methanol, $Zn_2(BzOip)_2(H_2O)\cdot0.22MeOH\cdot0.23H_2O$ from methanol, and $Co_2(BzOip)_2(H_2O)\cdot0.19H_2O$ from ethanol, $\lambda = 0.77490$ Å) or a Rigaku Mercury2 SCXMiniflex (Rigaku, Tokyo, Japan) diffractometer ($\lambda = 0.71075$ Å) at the University of St. Andrews. Appropriate scattering factors were applied using XDISP [35] within the WinGX suite [36]. Multi-scan absorption corrections were applied to the synchrotron and in-house crystallographic data using SADABS [37] and REQAB [38] programs, respectively. Data were solved using SHELXT [39] or SHELXS-97 [40] and refined on F^2 using SHELXL-97 [41] using the ShelXle User interface [42]. All non-hydrogen framework atoms were refined with anisotropic thermal displacement parameters, with equivalent disordered atoms constrained to have equal *Uij* values. All C-bound hydrogen atoms were included at their geometrically estimated position, while O-bound hydrogen atoms were located in the difference map and fixed at a distance of 0.90 Å from the oxygen atom to which they are bound. Pairs of hydrogen atoms belonging to a single water molecule were also fixed at a distance of 1.47(2) Å from each other. Disorder of the benzyl groups was modelled initially from the difference map, with each orientation having a common occupancy such that the sum of all occupancies were constrained to sum to 1.00. Once all orientations were modelled, a visual inspection was conducted to determine which orientations were mutually exclusive within the hexaphenyl unit, and all occupancies were then defined relative to a single free variable. Atoms belonging to partial occupancy solvent molecules were initially modeled with an isotropic thermal displacement parameter fixed at 0.08 and their occupancy allowed to refine. Visual inspection was then conducted to see which orientations of the benzyl group were too close for this solvent molecule to be present, and the O (and C) atoms given an occupancy related to the existing single free variable.

5. Conclusions

The level of disorder of the pendant benzyl groups in the frameworks $M_2(BzOip)_2(H_2O)\cdot$solvate (M = Co or Zn) was found to vary according to the polarity of the preparative solvent system used to produce the crystals, without changing the overall topology of the coordination framework. This demonstrates the more subtle ways that synthetic solvent choice can influence coordination polymer synthesis, even when the overall topology of the framework is not affected.

Supplementary Materials: The supplementary materials are available online at www.mdpi.com/2073-4352/8/1/6/s1. CCDC 1525893-1525897, 1526369-1526371, and 1526372-1526381 contain the supplementary crystallographic data for this paper. These data can be obtained free of charge from The Cambridge Crystallographic Data Centre via www.ccdc.cam.ac.uk/data_request/cif.

Acknowledgments: This work was funded by the British Heart Foundation (NH/11/8/29253) and the EPSRC (EP/K005499/1 and EP/K039210/1). This research used resources of the Advanced Light Source, which is a DOE Office of Science User Facility under contract no. DE-AC02-05CH11231.

Author Contributions: Laura J. McCormick conducted the synthetic work, crystallographic (in house) data collections and refinement, and prepared the manuscript. Samuel A. Morris performed the data collections at the ALS. Alexandra M. Z. Slawin, and Simon J. Teat secured funding for, and maintained the diffractometers at the University of St. Andrews and the Advanced Light Source, respectively. Russell E. Morris secured funding for the synthetic and operational aspects of this work and defined the initial project direction.

Conflicts of Interest: The authors declare no conflict of interest.

References

1. Cavka, J.H.; Jakobsen, S.; Olsbye, U.; Guillou, N.; Lamberti, C.; Bordiga, S.; Lillerud, K.P. A New Zirconium Inorganic Building Brick Forming Metal Organic Frameworks with Exceptional Stability. *J. Am. Chem. Soc.* **2008**, *130*, 13850–13851. [CrossRef] [PubMed]

2. Serre, C.; Millange, F.; Thouvenot, C.; Nogues, M.; Marsolier, G.; Louer, D.; Ferey, G. Very Large Breathing Effect in the First Nanoporous Chromium(III)-Based Solids: MIL-53 or $Cr^{III}(OH)\cdot\{O_2C-C_6H_4-CO_2\}\cdot\{HO_2C-C_6H_4-CO_2H\}_x\cdot H_2O_y$. *J. Am. Chem. Soc.* **2002**, *124*, 13519–13526. [CrossRef] [PubMed]

3. Ferey, G.; Mellot-Draznieks, C.; Serre, C.; Millange, F.; Dutour, J.; Surble, S.; Margiolaki, I. A Chromium Terephthalate-Based Solid with Unusually Large Pore Volumes and Surface Area. *Science* **2005**, *309*, 2040–2042. [CrossRef] [PubMed]

4. Li, H.; Eddaoudi, M.; O'Keeffe, M.; Yaghi, O.M. Design and synthesis of an exceptionally stable and highly porous metal-organic framework. *Nature* **1999**, *402*, 276–279. [CrossRef]

5. Dietzel, P.D.C.; Blom, R.; Fjellvåg, H. Base-Induced Formation of Two Magnesium Metal-Organic Framework Compounds with a Bifunctional Tetratopic Ligand. *Eur. J. Inorg. Chem.* **2008**, *2008*, 3624–3632. [CrossRef]

6. Gao, Q.; Jiang, F.-L.; Wu, M.-Y.; Huang, Y.-G.; Wei, W.; Zhang, Q.-F.; Hong, M.-C. Crystal Structures, Topological Analyses, and Magnetic Properties of Manganese-Dihydroxyterephthalate Complexes. *Aust. J Chem.* **2010**, *63*, 286–292. [CrossRef]

7. Bloch, E.D.; Murray, L.J.; Queen, W.L.; Chavan, S.; Maximoff, S.N.; Bigi, J.P.; Krishna, R.; Peterson, V.K.; Grandjean, F.; Long, G.J.; et al. Selective Binding of O_2 over N_2 in a Redox–Active Metal–Organic Framework with Open Iron(II) Coordination Sites. *J. Am. Chem. Soc.* **2011**, *133*, 14814–14822. [CrossRef] [PubMed]

8. Dietzel, P.D.C.; Morita, Y.; Blom, R.; Fjellvåg, H. An In Situ High-Temperature Single-Crystal Investigation of a Dehydrated Metal–Organic Framework Compound and Field-Induced Magnetization of One-Dimensional Metal–Oxygen Chains. *Angew. Chem. Int. Ed.* **2005**, *44*, 6354–6358. [CrossRef] [PubMed]

9. Dietzel, P.D.C.; Panella, B.; Hirscher, M.; Blom, R.; Fjellvåg, H. Hydrogen adsorption in a nickel based coordination polymer with open metal sites in the cylindrical cavities of the desolvated framework. *Chem. Commun.* **2006**, 959–961. [CrossRef] [PubMed]

10. Sanz, R.; Martínez, F.; Orcajo, G.; Wojtas, L.; Briones, D. Synthesis of a honeycomb-like Cu-based metal–organic framework and its carbon dioxide adsorption behavior. *Dalton Trans.* **2013**, *42*, 2392–2398. [CrossRef] [PubMed]

11. Rosi, N.L.; Kim, J.; Eddaoudi, M.; Chen, B.; O'Keeffe, M.; Yaghi, O.M. Rod Packings and Metal−Organic Frameworks Constructed from Rod-Shaped Secondary Building Units. *J. Am. Chem. Soc.* **2005**, *127*, 1504–1518. [CrossRef] [PubMed]

12. Eddaoudi, M.; Kim, J.; Rosi, N.; Vodak, D.; Wachter, J.; O'Keeffe, M.; Yaghi, O.M. Systematic Design of Pore Size and Functionality in Isoreticular MOFs and Their Application in Methane Storage. *Science* **2002**, *295*, 469–472. [CrossRef] [PubMed]

13. Silva, C.G.; Luz, I.; Llabres i Xamena, F.X.; Corma, A.; Garcia, H. Water Stable Zr–Benzenedicarboxylate Metal–Organic Frameworks as Photocatalysts for Hydrogen Generation. *Chem. Eur. J.* **2010**, *16*, 11133–11138. [CrossRef] [PubMed]

14. Ahnfeldt, T.; Gunzelmann, D.; Loiseau, T.; Hirsemann, D.; Senker, J.; Ferey, G.; Stock, N. Synthesis and Modification of a Functionalized 3D Open-Framework Structure with MIL-53 Topology. *Inorg. Chem.* **2009**, *48*, 3057–3064. [CrossRef] [PubMed]

15. Biswas, S.; Couck, S.; Grzywa, M.; Denayer, J.F.M.; Volkmer, D.; Van Der Voort, P. Vanadium Analogues of Nonfunctionalized and Amino-Functionalized MOFs with MIL-101 Topology–Synthesis, Characterization, and Gas Sorption Properties. *Eur. J. Inorg. Chem.* **2012**, *2012*, 2481–2486. [CrossRef]

16. Chui, S.S.-Y.; Lo, S.M.-F.; Charmant, J.P.H.; Orpen, A.G.; Williams, I.D. A Chemically Functionalizable Nanoporous Material [Cu$_3$(TMA)$_2$(H$_2$O)$_3$]$_n$. *Science* **1999**, *283*, 1148–1150. [CrossRef] [PubMed]

17. Ferey, G.; Serre, C.; Mellot-Draznieks, C.; Millange, F.; Surble, S.; Dutour, J.; Margiolaki, I. A Hybrid Solid with Giant Pores Prepared by a Combination of Targeted Chemistry, Simulation, and Powder Diffraction. *Angew. Chem. Int. Ed.* **2004**, *43*, 6296–6301. [CrossRef] [PubMed]

18. Horcajada, P.; Surble, S.; Serre, C.; Hong, D.-Y.; Seo, Y.-K.; Chang, J.-S.; Greneche, J.-M.; Margiolaki, I.; Ferey, G. Synthesis and catalytic properties of MIL-100(Fe), an iron(III) carboxylate with large pores. *Chem. Commun.* **2007**, 2820–2822. [CrossRef] [PubMed]

19. McCormick, L.J.; Morris, S.A.; Slawin, A.M.Z.; Teat, S.J.; Morris, R.E. Coordination Polymers of 5-Alkoxy Isophthalic Acids. *Cryst. Growth Des.* **2016**, *16*, 5771–5780. [CrossRef]

20. Yang, H.-B.; Northrop, B.H.; Zheng, Y.-R.; Ghosh, K.; Stang, P.J. Facile Self-Assembly of Neutral Dendritic Metallocycles via Oxygen-to-Platinum Coordination. *J. Org. Chem.* **2009**, *74*, 7067–7074. [CrossRef] [PubMed]

21. Yang, J.-X.; Zhang, X.; Cheng, J.-K.; Yao, Y.-G. A novel 1D→2D interdigitated framework directed by hydrogen bonds. *J. Mol. Struct.* **2011**, *991*, 31–34. [CrossRef]

22. Gole, B.; Bar, A.K.; Mukherjee, P.S. Modification of Extended Open Frameworks with Fluorescent Tags for Sensing Explosives: Competition between Size Selectivity and Electron Deficiency. *Chem. Eur. J.* **2014**, *20*, 2276–2291. [CrossRef] [PubMed]

23. Zhang, D.-W.; Zhao, G.-Y. Synthesis, Crystal Structure, and Photoluminescent Property of a New 1D→2D Interdigitated Framework. *Synth. React. Inorg. Met. Org. Nano-Met. Chem.* **2015**, *45*, 524–526. [CrossRef]

24. Su, Y.; Li, X.; Li, X.; Pan, H.; Wang, R. Effects of hydroxy substituents on Cu(II) coordination polymers based on 5-hydroxyisophthalate derivatives and 1,4-bis(2-methylimidazol-1-yl)benzene. *CrystEngComm* **2015**, *17*, 4883–4894. [CrossRef]

25. Gole, B.; Bar, A.K.; Mukherjee, P.S. Multicomponent Assembly of Fluorescent-Tag Functionalized Ligands in Metal–Organic Frameworks for Sensing Explosives. *Chem. Eur. J.* **2014**, *20*, 13321–13336. [CrossRef] [PubMed]

26. Li, X.; Li, J.; Li, M.-K.; Fei, Z. Synthesis, structures and photocatalytic properties of two new Co(II) coordination polymers based on 5-(benzyloxy)isophthalate ligand. *J. Mol. Struct.* **2014**, *1059*, 294–298. [CrossRef]

27. Zhang, X.; Yang, J.-X.; Yao, Y.-G. Tail of the Organic Ligand Templated Metal-Organic Framework. *J. Inorg. Organomet. Polym. Mater.* **2012**, *22*, 1189–1193. [CrossRef]

28. Perry, J.J.; McManus, G.J.; Zaworotko, M.J. Sextuplet phenyl embrace in a metal–organic Kagomé lattice. *Chem. Commun.* **2004**, 2534–2535. [CrossRef] [PubMed]

29. Dance, I.; Scudder, M. The Sextuple Phenyl Embrace, a Ubiquitous Concerted Supramolecular Motif. *J. Chem. Soc. Chem. Commun.* **1995**, 1039–1040. [CrossRef]

30. Dance, I.; Scudder, M. Supramolecular Motifs: Concerted Multiple Phenyl Embraces between Ph4P+ Cations Are Attractive and Ubiquitous. *Chem. Eur. J.* **1996**, *2*, 481–486. [CrossRef] [PubMed]

31. Sun, X.-F.; You, J.-M.; Xu, X.-H.; Li, X.-J. A Zn(II) Metal-organic Framework Consisting of Hexagonal Cavities Constructed by Benzyl-functionalized 5-Aminoisophthalate. *Chin. J. Struct. Chem.* **2015**, *34*, 562–568. [CrossRef]

32. Gupta, M.; Ahmad, M.; Singh, R.; Mishra, R.; Sahu, J.; Gupta, A.K. Zn(II)/Cd(II) based coordination polymers synthesized from a semi-flexible dicarboxylate ligand and their emission studies. *Polyhedron* **2015**, *101*, 86–92. [CrossRef]

33. Chucholowski, A.; Fingerle, J.; Iberg, N.; Marki, H.P.; Muller, R.; Pech, M.; Rouge, M.; Schmid, G.; Tschopp, T.; Wessel, H.P. Sulfuric Acid Esters of Sugar Alcohols. U.S. Patent US5,521,160, 28 May 1996.

34. Gibson, H.W.; Wang, H.; Niu, Z.; Slebodnick, C.; Zhakharov, L.N.; Rheingold, A.L. Rotaxanes from Tetraclams. *Macromolecules* **2012**, *45*, 1270–1280. [CrossRef]

35. Kissel, L.; Pratt, R.H. XDISP Program in WinGX. *Acta Crystallogr.* **1990**, *A46*, 170–175. [CrossRef]

36. Farrugia, L.J. *WinGX* and *ORTEP* for *Windows*: An update. *J. Appl. Crystallogr.* **2012**, *45*, 849–854. [CrossRef]

37. Krause, L.; Herbst-Irmer, R.; Sheldrick, G.M.; Stalke, D. Comparison of silver and molybdenum microfocus X-ray sources for single-crystal structure determination. *J. Appl. Crystallogr.* **2015**, *48*, 3–10. [CrossRef] [PubMed]

38. *CrystalClear-SM Expert*, version 2.1; Rigaku Americas: The Woodlands, TX, USA, 2015.

39. Sheldrick, G.M. *SHELXT*—Integrated space group and crystal-structure determination. *Acta Crystallogr.* **2015**, *A71*, 3–8. [CrossRef] [PubMed]

40. Sheldrick, G.M. A short history of *SHELX*. *Acta Crystallogr. Sect. A* **2008**, *64*, 112–122. [CrossRef] [PubMed]

41. Sheldrick, G.M. Crystal structure refinement with *SHELXL*. *Acta Crystallogr.* **2015**, *C71*, 3–8. [CrossRef]

42. Hübschle, C.B.; Sheldrick, G.M.; Dittrich, B. ShelXle: A Qt user interface for *SHELXL*. *J. Appl. Cryst.* **2011**, *44*, 1281–1284. [CrossRef] [PubMed]

crystals

MDPI

Article

A One-Dimensional Coordination Polymer Containing Cyclic [Ag4] Clusters Supported by a Hybrid Pyridine and Thioether Functionalized 1,2,3-Triazole

Shi-Qiang Bai * and Ivy Hoi Ka Wong

Institute of Materials Research and Engineering, A*STAR (Agency for Science, Technology and Research),
2 Fusionopolis Way, #08-03, Innovis, Singapore 138634, Singapore; wongi@imre.a-star.edu.sg
* Correspondence: bais@imre.a-star.edu.sg; Tel.: +65-6416-8966

Received: 7 December 2017; Accepted: 28 December 2017; Published: 2 January 2018

Abstract: A pyridine and thioether co-supported triazole ligand **L** (**L** = 2-((4-(3-(cyclopentylthio)propyl) -1H-1,2,3-triazol-1-yl)methyl)pyridine) has been synthesized using the CuAAC click reaction. This ligand supports the formation of a thermally stable, one-dimensional coordination polymer $[L_2Ag_4]_n \cdot 4n(BF_4)$ (**1**) possessing a cationic polymeric structure with $[Ag_4]$ metallomacrocycles, in which the ligand **L** displays chelate/bridging coordination modes using all four potential donors of nitrogen (N) and thioether (S). The dominant direction of the prism crystals of **1** aligns with the propagation of the chain in the lattice.

Keywords: hybrid ligand; click reaction; [Ag4] metallomacrocycle; coordination polymer; crystal morphology

1. Introduction

Coordination polymers are well-defined molecular materials consisting of metal centers and organic ligands, and have received considerable attention because of their fascinating crystalline structures and wide range of applications, including as luminescent sensors, catalysts, and porous and electrochemical materials [1–5]. The formation and dimensionality of coordination polymers are governed by the coordination number and geometry of the metal centers and the nature and arrangement of donor atoms on the ligand. Reaction conditions—such as temperature, solvent, metal-to-ligand ratio, and counter ions—also influence coordination diversity [6,7]. Many synthetic chemists have dedicated themselves to the construction of new organic ligands and coordination polymers, from which to discover the principles of controlled self-assembly. Significant effort has been devoted to the complementarity of metal precursor and rationally designed organic ligands [8–12]. Ag(I) with d^{10} electronic configuration, multiple coordination geometries and luminescence resulting from Ag$\cdots$Ag interactions is especially attractive to crystal engineers [13–17]. Ag(I) is a soft metal and so favored to N, O, and S donor ligands [18–21]. N-heterocyclic pyridine, pyrazole, imidazole, triazole, and multidentate ligands have been employed to support Ag(I) ions and clusters. We have a particular interest in hybrid ligands with different chemical donors that demonstrate hemilability for the design of new functional materials [22,23]. Schiff-base-, pyrazole-, and thioether-functionalized pyridines have been used to construct magnetic and luminescent coordination polymers of Mn^{II}, Ni^{II}, $Cu^{II/I}$, and Zn^{II} [24–30]. The copper-catalyzed azide-alkyne cycloaddition (CuAAC) reactions provide access to versatile, multidirectional 1,2,3-triazole ligands and we have used these ligands to make copper-iodide cluster-based coordination compounds and study the relationship between lattice weak interactions and crystal growth [31,32]. Different substituents on the ligands can be used to tune the possible coordination modes and these weak lattice interactions. The distance between two coordination sites (e.g., thioether (S) and pyridine–triazole NN groups) is also appropriate for stabilizing different sized copper-iodide clusters. As a continuation of this work, we herein report a new pyridine and thioether supported 1,2,3-triazole ligand L synthesized using the click reaction. This neutral ligand employs pyridine–triazole chelate-bridging and thioether bridging modes to stabilize Ag(I) centers and form a polymeric, cationic chain with cyclic [Ag$_4$] clusters.

2. Materials and Methods

All reagents were used as received. Infrared spectra were obtained on a PerkinElmer Spectrum 2000 FT-IR spectrometer from a sample in KBr disc. Elemental analyses were performed on a thermo electron corporation flash EA 1112 series analyzer. Electrospray ionization mass spectrometry (ESI-MS) was recorded in positive ion mode using a Shimadzu LCMS-IT-TOF mass spectrometer. UV–vis absorption spectrum was recorded on a Shimadzu UV-2501PC UV–vis recording spectrophotometer. Photoluminescence spectrum was measured on a Shimadzu RF-5301 PC spectrofluorophotometer. Powder X-ray diffraction data was collected on a Bruker D8 Advance X-ray diffractometer with Cu-Kα radiation (λ = 1.54056 Å). Thermogravimetric analysis (TGA) was carried out in an air stream using a *TA* Instruments TGA Q500 analyzer with a heating rate of 20 °C per min.

X-ray Crystallography

Single-crystal X-ray diffraction data were collected using a Bruker AXS SMART APEXII CCD diffractometer using Mo-Kα radiation (λ = 0.71073 Å). Data integration and scaling were performed using Bruker SAINT [33]. The empirical absorption correction was performed by SADABS [34]. The space group determination, structure solution, and least-squares refinements on $|F|^2$ were carried out with the Bruker SHELXL [35]. The structure was solved by direct method to locate the heavy atoms, followed by difference maps for the light non-hydrogen atoms. Anisotropic thermal parameters were refined for the rest of the non-hydrogen atoms. Hydrogen atoms were placed geometrically and refined isotropically. CCDC reference number: 1587041 (1). Crystal Data for C$_{32}$H$_{44}$Ag$_4$B$_4$F$_{16}$N$_8$S$_2$ (M = 1383.57 g/mol): monoclinic, space group P2$_1$/c, a = 15.6371(4) Å, b = 9.1234(3) Å, c = 16.6824(5) Å, β = 101.648(1)°, V = 2331.0(1) Å^3, Z = 2, T = 296(2) K, D$_{calc}$ = 1.971 g/cm^3, 39572 reflections measured (1.33° $\leq \Theta \leq$ 27.12°), 5147 unique (R$_{int}$ = 0.0277) which were used in all calculations. The final R$_1$ was 0.0733 (I > 2σ(I)) and wR$_2$ was 0.2093 (all data).

3. Results and Discussion

3.1. Synthesis

Sodium azide is potentially explosive. Only micro-scaled reactions should be performed. The alkyne precursor (4-pentyn-1-ylthio)-cyclopentane was prepared following literature procedures for thioether formation [28,32]. The crude alkyne was placed in a round bottom flask containing 2-(chloromethyl)pyridine hydrochloride (328 mg, 2 mmol), Na$_2$CO$_3$ (210 mg, 2 mmol), NaN$_3$ (156 mg, 2.5 mmol), CuI (23 mg, 0.12 mmol), and CH$_3$OH/H$_2$O (1:1 v:v, 6 mL) (Scheme 1). The reaction was stirred at 50 °C for 24 h. The residue was extracted with ethyl acetate (150 mL) and the organic layer was washed with water (3 × 20 mL), dried (with anhydrous Mg$_2$SO$_4$) and concentrated by rotary evaporator at 60 °C under vacuum. Column chromatography on silica gel with hexane/ethyl acetate (2:1 v:v) as eluent produced a band at R$_f$ = 0.1 that was collected and the solvent removed by rotary evaporator at 40 °C under vacuum. The product (L) was a pale yellow oil. Yield: 330 mg, 55%. 2-((4-(3-(Cyclopentylthio)propyl)-1H-1,2,3-triazol-1-yl)methyl)pyridine (L), (C$_{16}$H$_{22}$N$_4$S, MW 302.44). ESI-MS (m/z, %): [L+H]$^+$ (303, 100). ^{1}H NMR (CDCl$_3$, 500.2 MHz) δ: 8.56 (s, 1H, pyridine, NCH), 7.68–7.65(m, 1H), 7.45 (s, 1H, triazole), 7.25–7.23 (m, 1H), 7.16–7.13 (t, 1H), 5.61 (s, 2H, pyridine–CH$_2$–triazole), 3.06–3.01 (m, 1H, SCH), 2.81–2.78 (m, 2H), 2.56–2.52 (m, 2H), 1.97–1.91 (m, 4H), 1.69 (b, 2H), and 1.53–1.43 (m, 4H). ^{13}C NMR (CDCl$_3$, 125.8 MHz): 154.7, 149.5, 147.9, 137.6, 123.5,

122.5, and 121.6 (C in pyridine and triazole groups), 55.4 (pyridine–CH$_2$–triazole), 43.8 (SCH), 33.9, 31.2, 29.4, and 24.9.

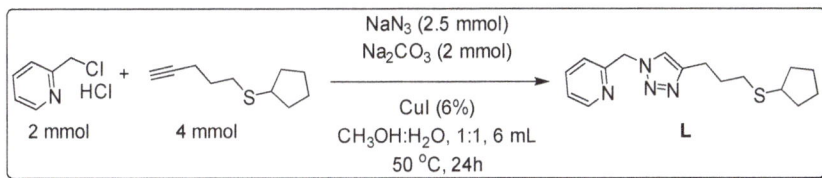

Scheme 1. Preparation of ligand **L**.

Complex **1** was prepared by mixing a methanol solution (5 mL) of ligand **L** (151 mg, 0.5 mmol) and a methanol solution (5 mL) of AgBF$_4$ (195 mg, 1 mmol). Single crystals were obtained by slow evaporation of the solvent over one week at room temperature (Scheme 2). The product was collected by filtration, washed with methanol and diethyl ether, and dried in vacuum oven for overnight at 60 °C. Yield: 270 mg, 78%. The powder sample of complex **1** is grey and no obvious solid-state luminescence was observed. Single crystals of **1** are prisms with a relative long axis. Anal. Calcd. for C$_{32}$H$_{44}$Ag$_4$B$_4$F$_{16}$N$_8$S$_2$ (1383.57): C, 27.78; H, 3.21; N, 8.10%. Found: C, 27.80; H, 3.27; N, 8.08%. Main IR bands (cm^{-1}): 3132(m), 3078(m), 2951(s), 2865(m), 1594(m), 1573(m), 1553(m), 1476(m), 1439(m), 1301(m), 1218(m), 1057(s, b, ν(BF$_4$$^-$), 756(m), 726(m), 596(m), 534(m), and 522(m). ESI-MS (*m/z*, %): [LAg]$^+$ (409, 100), [L$_2$Ag]$^+$ (713, 81).

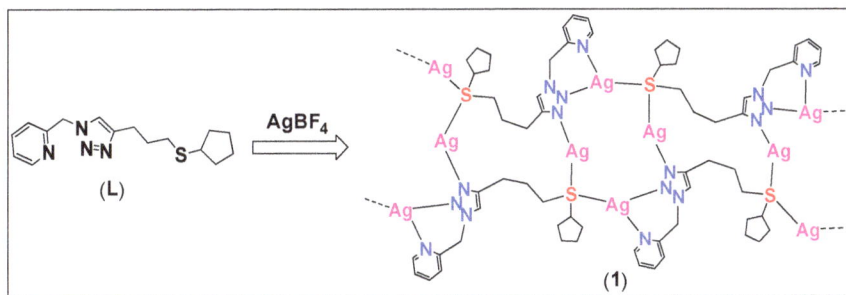

Scheme 2. Formation of cationic coordination polymer **1**.

3.2. UV–Vis and Photoluminescent Spectra of Ligand L

An ethanol solution of ligand **L** absorbed UV–vis light at 203 nm and between 240–280 nm with a maximum at about 260 nm (Figure 1a) and exhibited a broad emission between 400–500 nm with a maximum at 442 nm upon excitation at 358 nm. Neat liquid gave a blue emission under UV 365 nm light. (Figure 1b, insert) Coordination polymer **1** was synthesized by the reaction of **L** and AgBF$_4$ in methanol at room temperature (Scheme 2). IR spectroscopy indicated characteristic absorption by the BF$_4$$^-$ anion. The polymeric structure may be disassociated into small fragments (e.g., [LAg]$^+$ and [L$_2$Ag]$^+$) in solution which were detected by ESI-MS.

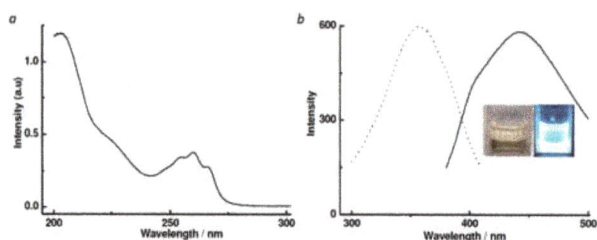

Figure 1. UV–vis (**a**) and excitation (dotted line) and emission (solid line) (**b**) spectra of ligand **L** in ethanol (0.1 M). (The inset images are ligand **L** in a quartz cuvette under normal and 365 nm lights, respectively).

3.3. Molecular Structure of Complex 1

Coordination polymer **1** crystallized in a monoclinic crystal system with space group $P\,2_1/c$. There is one neutral ligand **L**, two cationic Ag(I) centers and two lattice BF_4^- anions in the crystallographic asymmetric unit. (Figure 2a) Ag1 and Ag2 possess three-coordinated trigonal planer and two-coordinated linear geometries, respectively. Ag1 is surrounded by two chelating N donors from pyridine and triazole of ligand **L** and one bridging S donor from another ligand **L**. Ag2 is coordinated by one $3'$-$N_{triazole}$ of ligand **L** and one bridging S donor from another ligand **L**. These two S donors also bridge neighboring Ag1-Ag2 centers to form a $[Ag_4]$ metallomacrocycle. The cyclic $[Ag_4]$ units are further linked by double ligands and extend along the short *b* axis, which aligns with the dominant direction of the prism-shaped single crystals. (Figure 2c,d) The polymeric chains are parallel to each other in the lattice (Figure 2b). The counterions BF_4^- occupy the cavities created by the cationic polymeric chains. The Ag1$\cdots$Ag2 distances are 4.28 and 4.04 Å in the $[Ag_4]$ metallomacrocycle. The shortest Ag$\cdots$Ag distance of neighboring $[Ag_4]$ units is 6.06 Å. The shortest Ag$\cdots$Ag distance of neighboring chains is 7.60 Å.

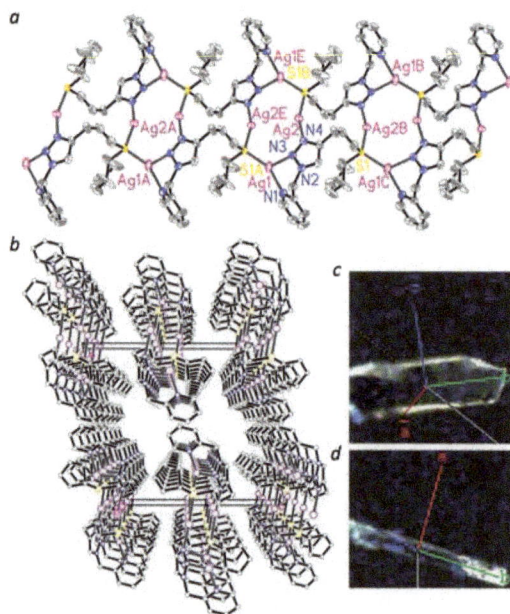

Figure 2. (**a**) Cationic and polymeric chain structure in **1**. (**b**) Packing structure in **1**. (BF_4^- anions are removed for clarity) (**c,d**) Single crystal indexes in different orientations of **1**.

3.4. Powder XRD and TGA

The experimental powder X-ray diffraction pattern for complex **1** showed good agreement with its simulated pattern determined from the single-crystal XRD experiment, supporting its phase purity (Figure 3). Thermogravimetric analysis (TGA) of **1** was conducted from room temperature to 900 °C under an air flow with a heating rate of 20 °C per min (Figure 4). Complex **1** is stable to about 230 °C. There are two weight loss stages between 230 and 890 °C, which correspond to the removal of ligand **L**, decomposition of BF$_4$ anion, and the formation of metallic silver with a residue amount of 31.5% (calcd. 31.2%).

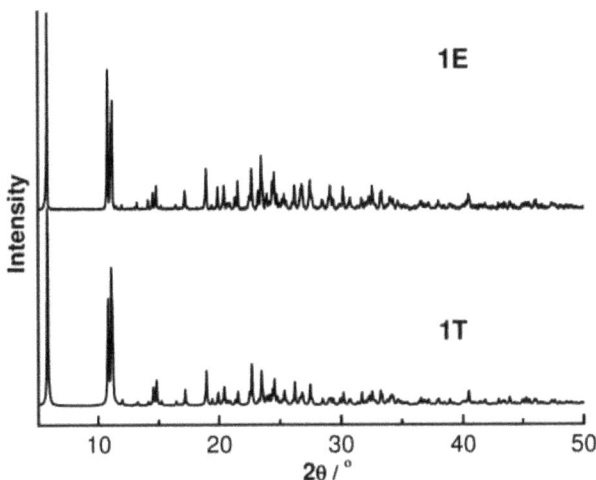

Figure 3. Powder XRD patterns of **1**. (T = theoretical profile referenced to the experimentally determined single-crystal XRD pattern; E = experimental data).

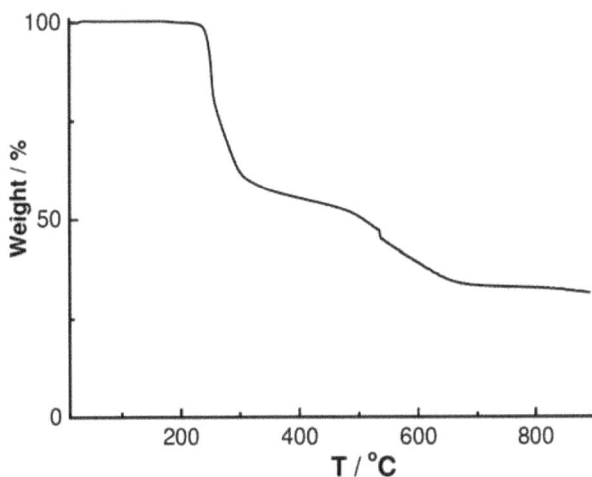

Figure 4. TGA curves of **1**.

4. Conclusions

In summary, a pyridine and thioether functionalized 1,2,3-triazole ligand with a relatively long methylene bridge (($CH_2)_3$) between NN and S donors was synthesized using the CuAAC click reaction. The reaction of ligand **L** and $AgBF_4$ afforded a novel, thermally stable polymeric chain structure with [Ag_4] metallomacrocyclics. The propagation of these chains in the lattice aligns with the dominant growth direction of the prism crystals. The observation of two types of Ag(I) coordination geometries (linear and trigonal planar) further emphasizes the diversity of Ag(I) coordination chemistry.

Acknowledgments: We thank Professor David J. Young of University of the Sunshine Coast of Australia for helpful discussion regarding this manuscript. We gratefully acknowledge the funding support (SERC Grant No. 1527200020) from the Institute of Materials Research and Engineering, A*STAR Singapore.

Author Contributions: Shiqiang Bai conceived and designed the experiments; Shiqiang Bai and Hoi Ka Ivy Wong. performed the experiments; Shiqiang Bai wrote the paper.

Conflicts of Interest: The authors declare no conflict of interest.

References

1. Cui, J.-W.; Hou, S.-X.; Hecke, K.V.; Cui, G.-H. Rigid versus semi-rigid bis(imidazole) ligands in the assembly of two Co.(II) coordination polymers: Structural variability, electrochemical properties and photocatalytic behavior. *Dalton Trans.* **2017**, *46*, 2892–2903. [CrossRef] [PubMed]
2. Pettinari, C.; Tăbăcaru, A.; Galli, S. Coordination polymers and metal–organic frameworks based on poly(pyrazole)-containing ligands. *Coord. Chem. Rev.* **2016**, *307*, 1–31. [CrossRef]
3. Zhang, X.; Wang, W.; Hu, Z.; Wang, G.; Uvdal, K. Coordination polymers for energy transfer: Preparations, properties, sensing applications, and perspectives. *Coord. Chem. Rev.* **2015**, *284*, 206–235. [CrossRef]
4. Hasegawa, Y.; Nakanishi, T. Luminescent lanthanide coordination polymers for photonic applications. *RSC Adv.* **2015**, *5*, 338–353. [CrossRef]
5. Zhang, W.-X.; Liao, P.-Q.; Lin, R.-B.; Wei, Y.-S.; Zeng, M.-H.; Chen, X.-M. Metal cluster-based functional porous coordination polymers. *Coord. Chem. Rev.* **2015**, *293*, 263–278. [CrossRef]
6. Tiekink, E.R.T.; Henderson, W. Coordination chemistry of 3- and 4-mercaptobenzoate ligands: Versatile hydrogen-bonding isomers of the thiosalicylate (2-mercaptobenzoate) ligand. *Coord. Chem. Rev.* **2017**, *341*, 19–52. [CrossRef]
7. Zhao, D.; Timmons, D.J.; Yuan, D.; Zhou, H.-C. Tuning the topology and functionality of metal-organic frameworks by ligand design. *Acc. Chem. Res.* **2011**, *44*, 123–133. [CrossRef] [PubMed]
8. Hao, J.-M.; Yu, B.-Y.; Van Hecke, K.; Cui, G.-H. A series of d^{10} metal coordination polymers based on a flexible bis(2-methylbenzimidazole) ligand and different carboxylates: Synthesis, structures, photoluminescence and catalytic properties. *Cryst. Eng. Commun.* **2015**, *17*, 2279–2293. [CrossRef]
9. Khlobystov, A.N.; Blake, A.J.; Champness, N.R.; Lemenovskii, D.A.; Majouga, A.G.; Zyk, N.V.; Schröder, M. Supramolecular design of one-dimensional coordination polymers based on silver(I) complexes of aromatic nitrogen-donor ligands. *Coord. Chem. Rev.* **2001**, *222*, 155–192. [CrossRef]
10. Liu, C.-Y.; Xu, L.-Y.; Ren, Z.-G.; Wang, H.-F.; Lang, J.-P. Assembly of silver(i)/n,n-bis(diphenylphosphanylmethyl) -3-aminopyridine/halide or pseudohalide complexes for efficient photocatalytic degradation of organic dyes in water. *Cryst. Growth Des.* **2017**, *17*, 4826–4834. [CrossRef]
11. Beheshti, A.; Soleymani-Babadi, S.; Mayer, P.; Abrahams, C.T.; Motamedi, H.; Trzybiński, D.; Wozniak, K. Design, synthesis, and antibacterial assessment of silver(i)-based coordination polymers with variable counterions and unprecedented structures by the tuning spacer length and binding mode of flexible bis(imidazole-2-thiones) ligands. *Cryst. Growth Des.* **2017**, *17*, 5249–5262. [CrossRef]
12. Dey, A.; Biradha, K. Anion and guest directed tetracyclic macrocycles of ag₅l₄ and ag₆l₄ with an arc-shaped ligand containing pyridine and benzimidazole units: reversal of anion selectivity by guest. *Cryst. Growth Des.* **2017**, *17*, 5629–5633. [CrossRef]
13. Reger, D.L.; Semeniuc, R.F.; Smith, M.D. Supramolecular architecture of a silver(i) coordination polymer supported by a new ligand containing four tris(pyrazolyl)methane units. *Inorg. Chem.* **2001**, *40*, 6545–6546. [CrossRef] [PubMed]

14. Han, L.; Wang, R.; Yuan, D.; Wu, B.; Lou, B.; Hong, M. Hierarchical assembly of a novel luminescent silver coordination framework with 4-(4-pyridylthiomethyl)benzoic acid. *J. Mol. Struct.* **2005**, *737*, 55–59. [CrossRef]

15. Chen, Z.-F.; Yu, L.-C.; Zhong, D.-C.; Liang, H.; Zhu, X.-H.; Zhou, Z.-Y. An unprecedented 1D ladder-like silver(I) coordination polymer with ciprofloxacin. *Inorg. Chem. Commun.* **2006**, *9*, 839–843. [CrossRef]

16. He, W.-J.; Luo, G.-G.; Wu, D.-L.; Liu, L.; Xia, J.-X.; Li, D.-X.; Dai, J.-C.; Xiao, Z.-J. An odd-numbered heptameric water cluster containing a puckered pentamer self-assembled in a Ag(I) polymeric solid. *Inorg. Chem. Commun.* **2012**, *18*, 4–7. [CrossRef]

17. Zorlu, Y.; Can, H.; Aksakal, F. A novel 2D chiral silver(I) coordination polymer assembled from 5-sulfosalicylic acid and (2S,4R)-4-hydroxyproline: Synthesis, crystal structure, HOMO–LUMO and NBO analysis. *J. Mol. Struct.* **2013**, *1049*, 368–376. [CrossRef]

18. Zhao, Y.-Q.; Fang, M.-X.; Xu, Z.-H.; Wang, X.-P.; Wang, S.-N.; Han, L.-L.; Li, X.-Y.; Sun, D. Construction of a crystalline Ag(I) coordination polymer based on a new ligand generated from unusual in situ aza-Michael addition reaction. *Cryst. Eng. Commun.* **2014**, *16*, 3015–3019. [CrossRef]

19. Yang, L.; Li, X.; Qin, C.; Wang, X.-L.; Shao, K.-Z.; Su, Z.-M. A highly electrical conducting, 3D supermolecular Ag(I) coordination polymer material with luminescent properties. *Inorg. Chem. Commun.* **2016**, *70*, 31–34. [CrossRef]

20. Bahemmat, S.; Ghassemzadeh, M.; Neumüller, B. A new silver(I) coordination polymer containing 3-methylthio-(1H)-1,2,4-triazole as precursor for preparation of silver nanorods. *Inorg. Chim. Acta*, **2015**, *435*, 159–166. [CrossRef]

21. Semitut, E.; Komarov, V.; Sukhikh, T.; Filatov, E.; Potapov, A. Synthesis, crystal structure and thermal stability of 1d linear silver(i) coordination polymers with 1,1,2,2-tetra(pyrazol-1-yl)ethane. *Crystals* **2016**, *6*, 138. [CrossRef]

22. Braunstein, P.; Naud, F. Hemilability of hybrid ligands and the coordination chemistry of oxazoline-based systems. *Angew. Chem. Int. Ed.* **2001**, *40*, 680–699. [CrossRef]

23. Deria, P.; Mondloch, J.E.; Karagiaridi, O.; Bury, W.; Hupp, J.T.; Farha, O.K. Beyond post-synthesis modification: Evolution of metal–organic frameworks via building block replacement. *Chem. Soc. Rev.* **2014**, *43*, 5896–5912. [CrossRef] [PubMed]

24. Bai, S.-Q.; Fang, C.-J.; He, Z.; Gao, E.-Q.; Yan, C.-H.; Hor, T.S.A. Chelating schiff base assisted azide-bridged Mn(II), Ni(II) and Cu(II) magnetic coordination polymers. *Dalton Trans.* **2012**, *41*, 13379–13387. [CrossRef] [PubMed]

25. Bai, S.-Q.; Leelasubcharoen, S.; Chen, X.; Koh, L.L.; Zuo, J.-L.; Hor, T.S.A. Crystallographic elucidation of chiral and helical Cu(II) polymers assembled from a heterodifunctional 1,2,3-triazole ligand. *Cryst. Growth Des.* **2010**, *10*, 1715–1720. [CrossRef]

26. Bai, S.-Q.; Kwang, J.Y.; Koh, L.L.; Young, D.J.; Hor, T.S.A. Functionalized 1,2,3-triazoles as building blocks for photoluminescent POLOs (polymers of oligomers) of copper(I). *Dalton Trans.* **2010**, *39*, 2631–2636. [CrossRef] [PubMed]

27. Jiang, L.; Wang, Z.; Bai, S.-Q.; Hor, T.S.A. "Click-and-click"—hybridised 1,2,3-triazoles supported Cu(I) coordination polymers for azide–alkyne cycloaddition. *Dalton Trans.* **2013**, *42*, 9437–9443. [CrossRef] [PubMed]

28. Bai, S.-Q.; Jiang, L.; Zuo, J.-L.; Hor, T.S.A. Hybrid NS ligands supported Cu(I)/(II) complexes for azide–alkyne cycloaddition reactions. *Dalton Trans.* **2013**, *42*, 11319–11326. [CrossRef] [PubMed]

29. Bai, S.-Q.; Jiang, L.; Sun, B.; Young, D.J.; Hor, T.S.A. Five Cu(I) and Zn(II) clusters and coordination polymers of 2-pyridyl-1,2,3-triazoles: Synthesis, structures and luminescence properties. *Cryst. Eng. Commun.* **2015**, *17*, 3305–3311. [CrossRef]

30. Bai, S.-Q.; Ke, K.L.; Young, D.J.; Hor, T.S.A. Structure and photoluminescence of cubane-like [Cu₄I₄] cluster-based 1D coordination polymer assembled with bis(triazole)pyridine ligand. *J. Organomet. Chem.* **2017**, 137–141. [CrossRef]

31. Bai, S.-Q.; Young, D.J.; Hor, T.S.A. Hydrogen-Bonding interactions in luminescent quinoline-triazoles with dominant 1d crystals. *Molecules* **2017**, *22*, 1600. [CrossRef] [PubMed]

32. Bai, S.-Q.; Jiang, L.; Tan, A.L.; Yeo, S.C.; Young, D.J.; Hor, T.S.A. Assembly of photoluminescent [Cu$_n$I$_n$] (n = 4, 6 and 8) clusters by clickable hybrid [N,S] ligands. *Inorg. Chem. Front.* **2015**, *2*, 1011–1018. [CrossRef]

33. Bruker AXS Inc. *SAINT Software Reference Manual*; Version 6.0; Bruker AXS Inc.: Madison, WI, USA, 2003.

34. Krause, L.; Herbst-Irmer, R.; Sheldrick, G. M.; Stalke, D. Comparison of silver and molybdenum microfocus X-ray sources for single-crystal structure determination. *J. Appl. Cryst.* **2015**, *48*, 3–10. [CrossRef] [PubMed]
35. Sheldrick, G.M. Crystal structure refinement with SHELXL. *Acta Cryst.* **2015**, *71*, 3–8.

crystals

MDPI

Article

Prussian Blue Analogue Mesoframes for Enhanced Aqueous Sodium-ion Storage

Huiyun Sun, Wei Zhang * and Ming Hu *

School of Physics and Materials Science, East China Normal University, Shanghai 200241, China;
Shyunyun16@163.com
* Correspondence: weizhang1990@163.com (W.Z.); mhu@phy.ecnu.edu.cn (M.H.); Tel.: +86-21-5434-5338

Received: 24 November 2017; Accepted: 3 January 2018; Published: 7 January 2018

Abstract: Mesostructure engineering is a potential avenue towards the property control of coordination polymers in addition to the traditional structure design on an atomic/molecular scale. Mesoframes, as a class of mesostructures, have short diffusion pathways for guest species and thus can be an ideal platform for fast storage of guest ions. We report a synthesis of Prussian Blue analogue mesoframes by top-down etching of cubic crystals. Scanning and transmission electron microscopy revealed that the surfaces of the cubic crystals were selectively removed by HCl, leaving the corners, edges, and the cores connected together. The mesoframes were used as a host for the reversible insertion of sodium ions with the help of electrochemistry. The electrochemical intercalation/de-intercalation of Na^+ ions in the mesoframes was highly reversible even at a high rate (166.7 C), suggesting that the mesoframes could be a promising cathode material for aqueous sodium ion batteries with excellent rate performance and cycling stability.

Keywords: Prussian Blue analogue; mesoframe; aqueous sodium ion battery

1. Introduction

Coordination polymers, for instance, metal-organic frameworks (MOFs) or porous coordination polymers (PCPs), are molecular solids assembled by metal ions/clusters and ligands [1–10]. The physical/chemical properties of coordination polymers, including the packing ordering of their component ions/molecules, are known to be determined by their structures [3,5,11,12]. Therefore, the most widely used approach for the property regulation of coordination polymers is structure design via manipulation of the atomic/molecular packing by intermolecular interactions [13–15]. Besides the great success achieved on the molecular/atomic scale, the structure design on a mesoscale (10~1000 nm) has recently been recognized as a promising choice [16–19]. By downsizing of a bulk crystal, the volume ratio of the surface units to the bulk units can be increased significantly, leading to a coupling effect from the bulk structure and the surface units [17,20,21]. For instance, the meso-sized coordination polymers can accommodate guest molecules still, but their phase-change type would be changed owing to the suppressing effect of the surface energy [14,22–24]. Downsizing can also shorten the diffusion distance when the guest molecules/ions travel inside, leading to high-rate adsorption/desorption ability.

Mesoframe represents a kind of mesostructures with open holes on each surface of a cube (Figure 1a). Comparing with the solid cubes, the open frames can provide larger accessible surfaces and shorter diffusion distances for guest species [25–36]. Such an ability of the mesoframes can be highly desirable for hosting guest alkaline ions. Very recently, Prussian Blue analogues $(A_xM[M'(CN)_6]_y \cdot V_{1-y} \cdot nH_2O$: A = alkali metal; M, M' = transition metals; V= M'(CN) vacancy; $0 \leq x \leq 2; 0 < y \leq 1)$, as a kind of coordination frameworks, have been recognized as an emerging class of energy storage materials because of their exceptional capability for hosting alkali metal ions

(Figure 1b) [37–46]. In particular, Prussian Blue analogues have interstitial sites that are quite suitable for the alkali metal ions with a larger radius than that of the Li^+ ions and thus can be used as electrodes for a new generation of Na-ion and K-ion batteries [47–57]. However, because of the poor electron conductivity of Prussian Blue analogues, it is necessary to make sure that Prussian Blue analogues can have a short diffusion distance for the guest ions [58–64]. Therefore, Prussian Blue analogues mesoframes are highly needed.

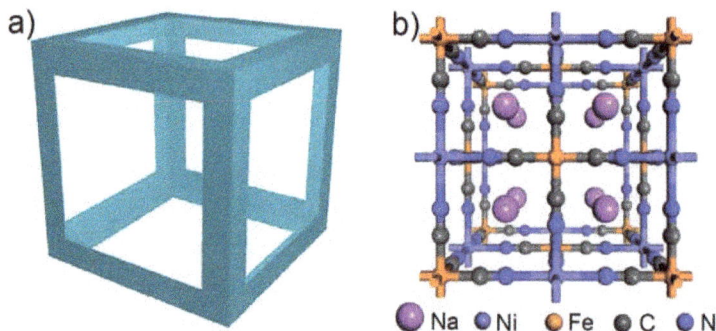

Na ● Ni ● Fe ● C ● N

Figure 1. (**a**) Scheme for Prussian Blue analogue mesoframe. (**b**) Schematic crystal structure for Prussian Blue analogue.

The most utilized strategy for the fabrication of mesoframes is selective etching, which occurs either during the synthesis or after the formation of compounds [14,22,65–73]. For example, when Prussian Blue crystal is crystallized in hot acid under a hydrothermal condition, the disassociation occurs during the growth process. When the disassociation rate is higher than the growth rate, etching of the crystal occurs [65]. By understanding the disassociation kinetics of different parts of a crystal, post-etching becomes possible. When the location of the defects is known, a hollow cavity can be constructed inside the crystals as desired [12]. In other cases, because of the spatial-dependent reactivity, for a cubic crystal of Prussian Blue analogues, the surfaces tend to be dissolved in the acidic/basic solution quicker than the corners and edges [22,74]. Although in situ etching methods have been extensively utilized [75–79], they are not suitable for investigating the differences between solid cubes and mesoframes because of the difficulties in controlling the size/shape and composition. Alternatively, post-etching provides opportunities for comparison of the solid cubes and mesoframes with similar size/shape and composition.

Despite increased knowledge of the positive effect of mesoframes, the role of mesoframes on alkaline ion intercalation/de-intercalation in aqueous environment remains unclear. Compared to the alkaline ion intercalation/de-intercalation in organic electrodes, aqueous alkaline ion storage is safer and cheaper and thus is an indispensable solution for grid-scale energy storage [80–84]. In this work, we fabricated Prussian Blue analogue ($Na_2Ni[Fe(CN)_6]$, **1**) mesoframes by post-etching the cubic crystals. The mesoframes of **1** showed improved Na^+ ion intercalation/de-intercalation at a high rate (166.7 C).

2. Materials and Methods

2.1. Materials

Sodium hexacyanoferrate ($Na_4[Fe(CN)_6] \cdot 10H_2O$), nickel chloride hexahydrate ($NiCl_2 \cdot 6H_2O$), hydrochloric acid (HCl), and trisodium citrate dihydrate ($C_6H_5Na_3O_7 \cdot 2H_2O$) were purchased from Sinopharm Chemical Reagent Co. and used without further purification.

2.2. Synthesis of Cubic Crystals of 1

Sodium hexacyanoferrate (5 mmol, 2.425 g) was dissolved in water (250 mL) to form a clear solution (Solution A). Nickel chloride hexahydrate (2.5 mmol, 587 mg) and trisodium citrate dihydrate (12.7 mmol, 3.75 g) were dissolved in water (250 mL) to form Solution B. Solutions A and B were mixed under magnetic stirring for 5 min and then allowed to age at 25 °C for 24 h. The product was obtained after centrifugation and washed with water and ethanol several times. Finally, the cubic crystals of **1** were collected by drying at 25 °C in a vacuum oven for 24 h.

2.3. Synthesis of Mesoframes of 1

The cubic crystals of **1** (50 mg) were added to 20 mL of 0.4 M HCl. The solution was ultrasonicated for 1 h. Then, the product was obtained after centrifugation and washed with water and ethanol several times. Finally, the mesoframes were collected by drying at 25 °C in vacuum oven for 24 h. We noted that Fe(II) in the surfaces of the crystals was easily oxidized into Fe(III) during etching in atmosphere according to the FT-IR spectrum (Figure S8). The existence of Fe(III) in the framework changed the crystal structure from *R-3* into *Fm-3m* (Figure S9). Therefore, the powder needed to be reduced electrochemically to convert the Fe(III) into Fe(II), making sure the crystals structure of the mesoframes was *R-3* before further characterization.

2.4. Characterization

The morphology of the products was observed by a field emission scanning electron microscope (FESEM, Hitachi S-4800) (Hitachi, Tokyo, Japan) and a transmission electron microscope (TEM, JEOL JEM-2100 F) ((JEOL, Tokyo, Japan). The powder X-ray diffraction patterns were measured with a Rigaku RINT 2500 X diffractometer with Cu Kα radiation (35 kV, 25 mA). The Fourier transform infrared spectroscopy (FT-IR) spectra were obtained with a Thermo Nicolet i50.

2.5. Electrochemical Measurement

The electrochemical performance was measured using three-electrode cells. The working electrode was prepared as follows: the powder of **1** (70%), acetylene black (20%), and poly(vinylidene difluoride) (10%) were well mixed in N-methyl-2-pyrrolidone until a homogeneous slurry was formed. The slurry was then coated onto the carbon cloth with an area of 1 cm × 2 cm and was dried at 80 °C in a vacuum oven for 15 h. The mass loading of the active materials was around 2 mg cm^{-2} for each test. An Ag/AgCl electrode was used as a reference. The counter electrode contained activated carbon instead of the prepared materials, and it was also coated onto the carbon cloth. These three electrodes were immersed in the neutral solution of 1 M NaNO$_3$ and measured with the CHI 660E (Shanghai Chen Hua Instrument, Shanghai, China) electrochemical workstation.

3. Results

The crystals of **1** were synthesized by a weak-chelation-agent-assisted crystallization method [85]. Approximately, nickel chloride was mixed with tri-sodium citrate to form a green solution at first. Then, the green solution was mixed with a sodium hexacyanoferrate solution. After aging at room temperature for 24 h, the green precipitates were harvested. Scanning electron microscopy (SEM) revealed that the obtained crystals were of a cubic shape with a size ranging from 200 to 400 nm (Figure 2a). Transmission electron microscopy confirmed that the crystals were solid (Figure 2b). The cubic crystals of **1** were immersed into an HCl solution (0.4 mol L^{-1}) under continuous stirring. After 1 h, the samples were harvested by centrifugation. SEM images illustrate that the center of the surfaces of the cubic crystals was empty (Figure 2c). The corners and edges were well-retained after etching. The TEM image implies that there were solid cores inside each particle (Figure 2d). All the evidence suggests that the mesoframes of **1** were successfully obtained after post-etching the cubic crystals. To facilitate our discussion, the cubic crystals of **1** are denoted as **1**$_{cube}$, while the mesoframes are designated as **1**$_{frame}$.

Figure 2. (**a**) SEM image of cubic crystals. (**b**) TEM image of cubic crystals. (**c**) SEM image of the product from etching of the cubic crystals. Inset is an enlarged image of a particle. (**d**) TEM image of the product from etching of the cubic crystals.

The crystal structures of **1**$_{cube}$ and **1**$_{frame}$ were investigated by powder X-ray diffraction (PXRD). The diffraction profiles of both samples are similar, indicating an identical crystal structure (Figure 3). The diffraction patterns can be assigned to a rhombohedral structure (*R*–3), which is typical for Prussian Blue analogues with the interstitial sites occupied by a rich amount of Na$^+$ ions [86]. The slight broadening of the peaks of **1**$_{frame}$ is probably due to the lattice defects induced by etching. FT-IR spectra are shown in Figure 4. For both samples, a similar peak near 2090 cm^{-1} was observed. This band belongs to the characteristic of stretching vibration of Fe–CN–Ni [87]. Therefore, the PXRD and FT-IR investigations confirm that both **1**$_{cube}$ and **1**$_{frame}$ present the same crystal structure and composition, allowing us to explore the mesostructural property.

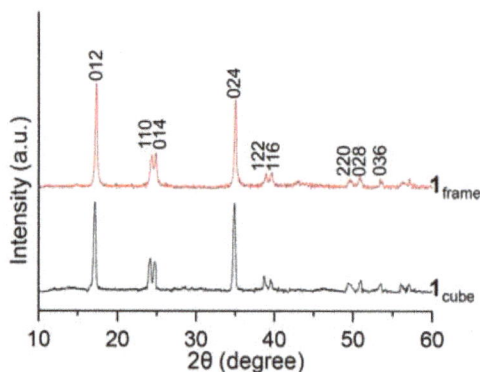

Figure 3. Powder X-ray diffraction (PXRD) patterns of **1** crystals.

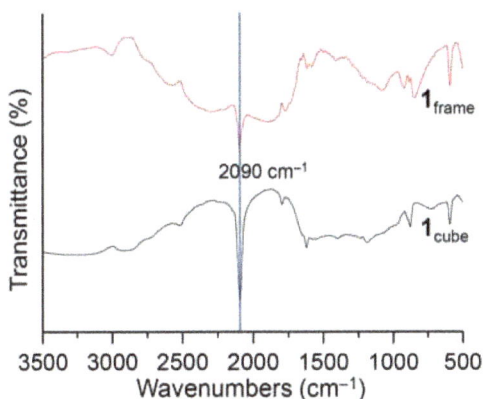

Figure 4. FT-IR spectra of **1** crystals.

To investigate the role of the frame-like mesostructure, we used the powders of **1** to host Na$^+$ ions. We selected an electrochemistry-controlled method to probe the Na-ion insertion/extraction amount. The crystals of **1** were used as cathodes in a three-electrode system in an NaNO$_3$ solution. In this aqueous battery system, a reversible insertion/extraction of Na$^+$ ions into/from the frameworks of **1** can be realized by electrochemical redox of Fe^{2+}/Fe^{3+} ions. For the storage/consumption of one electron, an Na$^+$ ion enters/leaves the host frameworks. Herein, we simply recorded the capacity change to understand the intercalation/de-intercalation of the Na$^+$ ion. The galvanostatic charge/discharge curves shown in Figure 5 present that **1**$_{frame}$ can deliver a reversible specific capacity of 61 mAh·g^{-1} at a current density of 70 mA·g^{-1} between 0.001 to 1.0 V (vs. Ag/AgCl). The discharge curve contains a plateau at 0.3–0.6 V (vs. Ag/AgCl), corresponding to the insertion of Na$^+$ ions. Moreover, **1**$_{cube}$ shows similar specific capacity (60 mAh·g^{-1}) at a current density of 70 mA·g^{-1} between 0.001 to 1.0 V (vs. Ag/AgCl). As we can see in the cyclic voltammetry (CV) curves (Figure 6), the center of the positions of redox peaks of the **1**$_{cube}$ crystals is at around 0.4 V (vs. Ag/AgCl), and the center of the peaks of **1**$_{frame}$ locates at around 0.36 V (vs. Ag/AgCl). The redox peaks of **1**$_{frame}$ are much sharper corresponding to a larger quantity of the intercalated Na$^+$ ion in the frameworks of **1**$_{frame}$ at this potential.

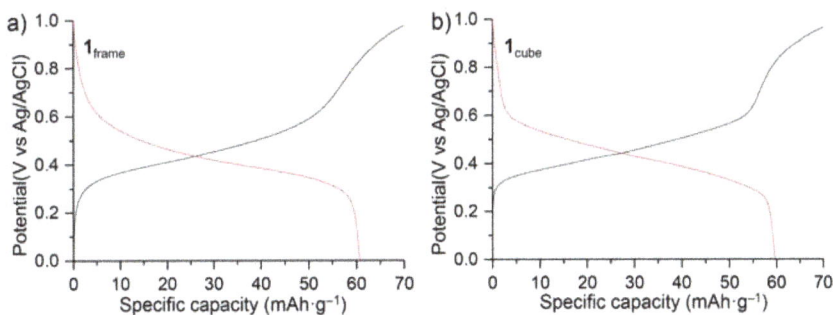

Figure 5. (**a,b**) Galvanostatic charge/discharge curves of **1**$_{frame}$ and **1**$_{cube}$ at a current density of 70 mA·g^{-1}, respectively.

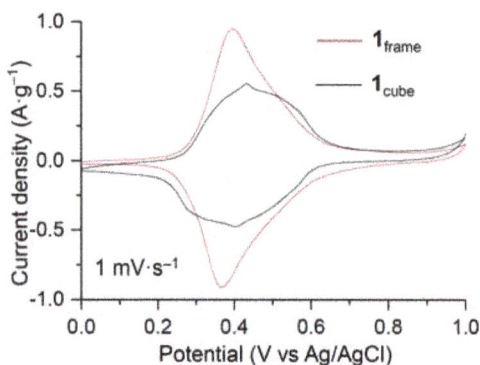

Figure 6. CV curves of **1** crystals at a scanning rate of 1 mV·s^{-1}.

We investigated the rate performance of the crystals of **1**. Figure 7 shows that **1**$_{frame}$ and **1**$_{cube}$ have similar capacities (approximately 60 mAh·g^{-1}) at a low current density 70 mA·g^{-1} (1 C). However, with the increase in the current rate, the capacity of **1**$_{cube}$ decreased more drastically than the **1**$_{frame}$. For instance, when the charge/discharge current density was as high as 10 A·g^{-1} (166.7 C), the capacity retention of **1**$_{frame}$ was as high as 88%. However, the **1**$_{cube}$ could only retain 41% of the initial capacity (Figure 7). The wonderful rate performance of **1**$_{frame}$ suggests that this material could be suitable for long-term cycling. Figure 8 illustrated that **1**$_{frame}$ could deliver a high capacity at a current density of 83.3 C (5 A·g^{-1}). The retained capacity is 94.34% of the initial capacity after 500 cycles (Figure 8). To check the performance at higher temperature, the **1**$_{frame}$ was measured at a different current density and cycled at 5 A·g^{-1} at 40 °C (Figure S1). The rate and cycling performances, compared with performances at 25 °C, were almost unchanged. This result indicated that the rate and cycling performances were not affected at 40 °C [88].

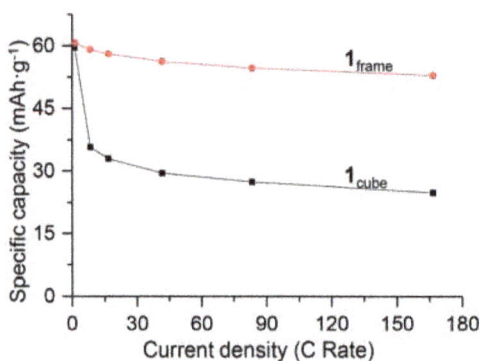

Figure 7. Rate performance of the **1** crystals.

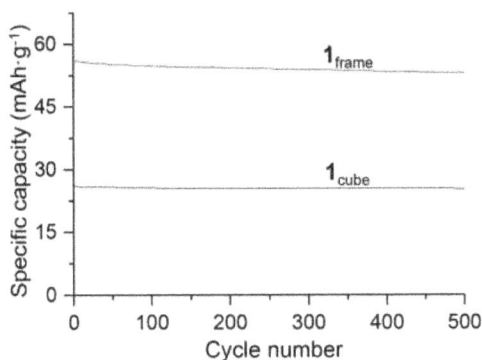

Figure 8. Cycling performance of the **1** crystals at a current density of 5 A·g^{-1}.

4. Discussion

On the basis of the relationship between the transferred electrons and the amount of the intercalated Na$^+$ ions, $\Delta N_e = \Delta N_{Na+}$, we converted the electrochemical data into the intercalation-deintercalation curves for Na$^+$ ions against the potential (vs. Ag/AgCl) (Figures 9 and 10). Apparently, both 1_{cube} and 1_{frame} can accommodate the same amount of Na$^+$ ions at a lower uptake rate, indicating that both samples have a similar capacity for the Na$^+$ ions (Figure 9). However, the uptake of Na$^+$ ions by 1_{cube} is much lower than that by 1_{frame} at a higher uptake rate (Figure 10). The uptake by 1_{frame} did not decrease too much, while 50% of the uptake by 1_{cube} was lost. Apparently, the interstitial sites in 1_{frame} could be utilized almost completely at a high rate, while the number of the accessed interstitial sites in 1_{cube} was significantly reduced.

There are three factors that may be affected by the meso-structure: (1) the contact between the crystal and the electrolyte; (2) the crystal structure and meso-structure change during the intercalation process; (3) the diffusion distance inside the frameworks. For Factor 1, we firstly measured the specific surface area of 1_{frame} and 1_{cube} by the Brunauer–Emmett–Teller (BET) method (Figure S2). The specific surface area of 1_{frame} (59.6 m^2·g^{-1}) is close to that of 1_{cube} (49.2 m^2·g^{-1}). Such a small difference can be ignored by considering that the crystals have to be well mixed with the conductors and electrolytes during the electrode preparation. Therefore, Factor 1 can be ignored. For Factor 2, we need to understand the structure change during the intercalation process. The ex situ PXRD analysis was carried out. For both crystals, a similar phase change between the R-3 and Fm-3m structures was recorded as shown in Figures S3 and S4. The structure change is due to the accommodation and release of Na$^+$ ions. The shape of the two samples was also checked by SEM after 500 cycles. We noted that the meso-structures of both samples were retained (Figure S5). It should be noted that the small particles are conductive carbons that are not crystals of **1**. This result suggests that Factor 2, for the mesoframes, is not different from the cube and thus should not be considered as the reason.

Regarding Factor 3, the 1_{frame} obviously has a shorter diffusion pathway for the Na$^+$ ions than that of 1_{cube} since various parts of the crystals have been removed. For the intercalation of Na$^+$ ions into Prussian Blue analogues, the solvated Na$^+$ ions need to first be de-solvated. Then, the de-solvated Na$^+$ ions can enter the frameworks of the host. The entered Na$^+$ ions may be partially re-solvated by the mobile water molecules in the crystals during the next diffusion stage. Because re-solvation can enlarge the radius of the Na$^+$ ions, the diffusion of the re-solvated Na$^+$ ions is slower than the de-solvated Na$^+$ ions. Shortening the diffusion distance can likely minimize the diffusion of the re-solvated Na$^+$ ions and accelerate the intercalation process, leading to the utilization of the interstitial sites adequately. To confirm this point, we measured the diffusion coefficient for both samples. The diffusion coefficient of Na$^+$ ions can be calculated with the Randles–Sevcik equation [89]: $I_p = 2.69 \times 10^5 n^{3/2} S C_{Na} D_{Na}^{1/2} v^{1/2}$, where I_p is the peak current (A), n is the number of electrons transferred per molecule during the intercalation/deintercalation procedure (n = 1 for Fe^{2+}/Fe^{3+} redox pair), S is the effective contact area between the electrode and electrolyte (here the area of the electrode = 1 cm^2), C_{Na} is the molar concentration of Na^+ ions (1.3×10^{-3} $mol \cdot cm^{-3}$) [90], D_{Na} is the diffusion coefficient of Na^+ ions ($cm^2 \cdot s^{-1}$), and v is the scan rate ($V \cdot s^{-1}$). We carried out CV scans at various scan rates from 0.2 to 1 mV s^{-1} (Figures S6a and S7a). The peak current correlates linearly to the square root of the scan rate for both samples. On the basis of the slopes received from the plots of the anodic I_p versus $v^{1/2}$ (Figures S6b and S7b), the calculated D_{Na} is 7.35×10^{-9} $cm^2 \cdot s^{-1}$ for 1_{cube}, and 14.0×10^{-9} $cm^2 \cdot s^{-1}$ for 1_{frame}, respectively. Apparently, the diffusion coefficient for Na^+ in 1_{frame} is significantly larger than in 1_{cube}. The diffusion coefficient strongly relates to the moving of the Na^+ ions, including the transfer from the electrolytes to the interstitial sites and the movement among the interstitial sites. In general, the transfer part is fast, while the moving part is slow. Because the transfer part for both samples is similar owing to their close specific surface areas, the less contribution there is from the moving part, the larger the diffusion coefficient can be. The moving step in 1_{frame} is shorter than in 1_{cube} because of the shape of the mesoframe, indicating less contribution to the diffusion coefficient from the moving step in 1_{frame}. Therefore, the diffusion coefficient for 1_{frame} is larger than for 1_{cube}. The larger diffusion coefficient means that the Na^+ ions can move fast and thus utilize the interstitial sites at a high current rate, leading to a larger capacity at a high current density. Therefore, we can say the short diffusion distance in 1_{frame} enlarges the D_{Na} and is thus the apparent reason for the better rate performance of the mesoframe structure.

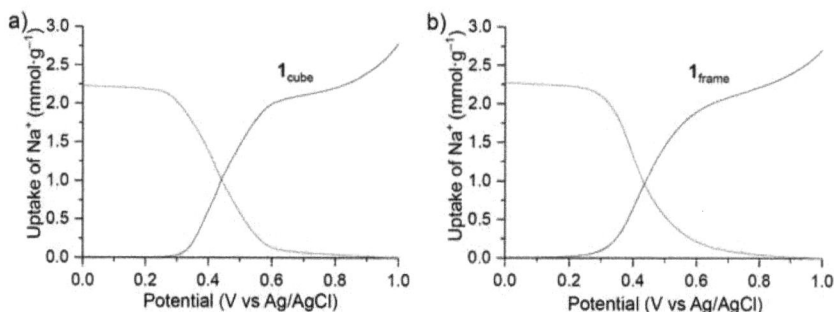

Figure 9. (**a**) Accommodation–release curves for Na^+ ions in 1_{cube} against the potential (vs. Ag/AgCl). (**b**) Accommodation–release curves for Na^+ ions in 1_{frame} against the potential (vs. Ag/AgCl).

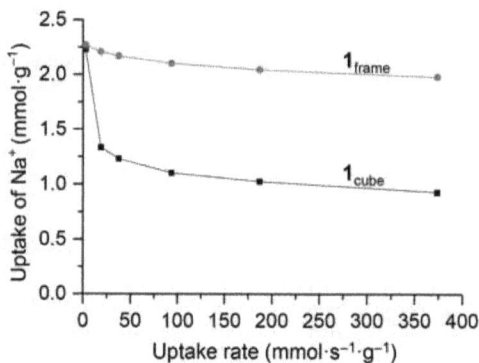

Figure 10. Uptake of Na^+ ions by **1** crystals at different uptake rates.

The above discussion suggests that the difference in Na-ion insertion/extraction at high rates for both samples can be explained as follows. At a low rate, the Na^+ ions can be inserted/extracted almost fully in/from both of the cubes and mesoframes. When the rate becomes high (for example, higher than 70 mA·g^{-1}), Na^+ ions may be hard to access to the center part of the cubes during the insertion process, while a large percentage of the interstitial site of the mesoframes can be accessed. Therefore, the mesoframes can have a much larger specific capacity, especially at a high current rate (higher than 0.07 A·g^{-1}), representing an excellent rate performance. In addition, as our cycling test was performed at a high current rate, the cycling stability of the mesoframes of course should be superior compared to the cubes as well.

5. Conclusions

In summary, we synthesized monocrystalline NiFe(II) Prussian Blue analogue mesoframes by a chemical etching method. This morphology significantly reduced the diffusion pathway for Na^+ ions inside the crystal, leading to fast guest ion intercalation/de-intercalation, which is hard to achieve in solid crystals. The fast guest ion accommodation ability of the crystals could be used in aqueous Na-ion batteries. Based on the mesoframes, cathodes suitable for a high-rate charging/discharging (166 C) aqueous Na-ion batteries were fabricated, and it was demonstrated that the mesoframe is a promising meso-architecture for an enhancement in the performance of coordination polymers. In addition, the impressive cycling performance of the mesoframes may be useful for grid-scale energy storage.

Supplementary Materials: The following are available online at www.mdpi.com/2073-4352/08/01/23/s1. Figure S1: Rate performance (**a**) and cycling performance (**b**) of **1**$_{frame}$ at different temperatures; Figure S2: N_2 adsorption-desorption isotherms of **1** crystals; Figure S3: Ex situ PXRD patterns of the electrochemically active cubic crystals at charge/discharge during the first cycle at a current density of 0.07A·g^{-1}; Figure S4: Ex situ PXRD patterns of the electrochemically active mesoframes at charge/discharge during the first cycle at a current density of 0.07A·g^{-1}; Figure S5: (**a**) SEM of cubic crystals after cycling. (**b**) SEM of mesoframes after cycling for 500 cycles; Figure S6: (**a**) Cyclic voltammetry curves of **1**$_{cube}$ at various scanning rates. (**b**) The anodic and cathodic peak currents as functions of the square root of scanning rates; Figure S7: (**a**) Cyclic voltammetry curves of **1**$_{frame}$ at various scanning rates. (**b**) The anodic and cathodic peak currents as functions of the square root of scanning rates; Figure S8: FT-IR spectrum of mesoframes before electrochemical activation; Figure S9: PXRD patterns of mesoframes before electrochemical activation.

Acknowledgments: This work was supported by the National Natural Science Foundation of China (Grant No. 21401056, 21473059) and the Large Instruments Open Foundation of East China Normal University.

Author Contributions: Ming Hu and Wei Zhang conceived and designed the experiments; Huiyun Sun performed the experiments; Huiyun Sun and Ming Hu analyzed the data; Ming Hu and Huiyun Sun wrote the paper.

Conflicts of Interest: The authors declare no conflict of interest.

References

1. James, S.L. Metal-organic frameworks. *Chem. Soc. Rev.* **2003**, *32*, 276–288. [CrossRef] [PubMed]
2. Rowsell, J.L.C.; Yaghi, O.M. Metal-organic frameworks: A new class of porous materials. *Micropor. Mesopor. Mater.* **2004**, *73*, 3–14. [CrossRef]
3. Eddaoudi, M.; Kim, J.; Rosi, N.; Vodak, D.; Wachter, J.; O'Keeffe, M.; Yaghi, O.M. Systematic design of pore size and functionality in isoreticular MOFs and their application in methane storage. *Science* **2002**, *295*, 469–472. [CrossRef] [PubMed]
4. Kitagawa, S.; Kitaura, R.; Noro, S. Functional porous coordination polymers. *Angew. Chem. Int. Ed.* **2004**, *43*, 2334–2375. [CrossRef] [PubMed]
5. Chui, S.S.Y.; Lo, S.M.F.; Charmant, J.P.H.; Orpen, A.G.; Williams, I.D. A chemically functionalizable nanoporous material [Cu$_3$(TMA)$_2$(H$_2$O)$_3$]$_n$. *Science* **1999**, *283*, 1148–1150. [CrossRef] [PubMed]
6. Hwang, Y.K.; Hong, D.-Y.; Chang, J.-S.; Jhung, S.H.; Seo, Y.-K.; Kim, J.; Vimont, A.; Daturi, M.; Serre, C.; Ferey, G. Amine grafting on coordinatively unsaturated metal centers of MOFs: Consequences for catalysis and metal encapsulation. *Angew. Chem. Int. Ed.* **2008**, *47*, 4144–4148. [CrossRef] [PubMed]
7. Dan-Hardi, M.; Serre, C.; Frot, T.; Rozes, L.; Maurin, G.; Sanchez, C.; Ferey, G. A new photoactive crystalline highly porous titanium(iv) dicarboxylate. *J. Am. Chem. Soc.* **2009**, *131*, 10857–10859. [CrossRef] [PubMed]

8. Furukawa, S.; Hirai, K.; Nakagawa, K.; Takashima, Y.; Matsuda, R.; Tsuruoka, T.; Kondo, M.; Haruki, R.; Tanaka, D.; Sakamoto, H.; et al. Heterogeneously hybridized porous coordination polymer crystals: fabrication of heterometallic core-shell single crystals with an in-plane rotational epitaxial relationship. *Angew. Chem. Int. Ed.* **2009**, *48*, 1766–1770. [CrossRef] [PubMed]

9. Salles, F.; Maurin, G.; Serre, C.; Llewellyn, P.L.; Knoefel, C.; Choi, H.J.; Filinchuk, Y.; Oliviero, L.; Vimont, A.; Long, J.R.; et al. Multistep n-2 breathing in the metal-organic framework co(1,4-benzenedipyrazolate). *J. Am. Chem. Soc.* **2010**, *132*, 13782–13788. [CrossRef] [PubMed]

10. Furukawa, S.; Reboul, J.; Diring, S.; Sumida, K.; Kitagawa, S. Structuring of metal-organic frameworks at the mesoscopic/macroscopic scale. *Chem. Soc. Rev.* **2014**, *43*, 5700–5734. [CrossRef] [PubMed]

11. Rungtaweevoranit, B.; Zhao, Y.; Choi, K.M.; Yaghi, O.M. Cooperative effects at the interface of nanocrystalline metal-organic frameworks. *Nano Res.* **2016**, *9*, 47–58. [CrossRef]

12. Hu, M.; Belik, A.A.; Imura, M.; Yamauchi, Y. Tailored design of multiple nanoarchitectures in metal-cyanide hybrid coordination polymers. *J. Am. Chem. Soc.* **2013**, *135*, 384–391. [CrossRef] [PubMed]

13. Huo, J.; Wang, L.; Irran, E.; Yu, H.; Gao, J.; Fan, D.; Li, B.; Wang, J.; Ding, W.; Amin, A.M.; et al. Hollow ferrocenyl coordination polymer microspheres with micropores in shells prepared by ostwald ripening. *Angew. Chem. Int. Ed.* **2010**, *49*, 9237–9241. [CrossRef] [PubMed]

14. Ameloot, R.; Vermoortele, F.; Vanhove, W.; Roeffaers, M.B.J.; Sels, B.F.; De Vos, D.E. Interfacial synthesis of hollow metal-organic framework capsules demonstrating selective permeability. *Nat. Chem.* **2011**, *3*, 382–387. [CrossRef] [PubMed]

15. Brinzei, D.; Catala, L.; Mathoniere, C.; Wernsdorfer, W.; Gloter, A.; Stephan, O.; Mallah, T. Photoinduced superparamagnetism in trimetallic coordination nanoparticles. *J. Am. Chem. Soc.* **2007**, *129*, 3778–3779. [CrossRef] [PubMed]

16. Volatron, F.; Catala, L.; Riviere, E.; Gloter, A.; Stephan, O.; Mallah, T. Spin-crossover coordination nanoparticles. *Inorg. Chem.* **2008**, *47*, 6584–6586. [CrossRef] [PubMed]

17. Catala, L.; Brinzei, D.; Prado, Y.; Gloter, A.; Stephan, O.; Rogez, G.; Mallah, T. Core-multishell magnetic coordination nanoparticles: Toward multifunctionality on the nanoscale. *Angew. Chem. Int. Ed.* **2009**, *48*, 183–187. [CrossRef] [PubMed]

18. Diring, S.; Furukawa, S.; Takashima, Y.; Tsuruoka, T.; Kitagawa, S. Controlled multiscale synthesis of porous coordination polymer in nano/micro regimes. *Chem. Mater.* **2010**, *22*, 4531–4538. [CrossRef]

19. Doherty, C.M.; Buso, D.; Hill, A.J.; Furukawa, S.; Kitagawa, S.; Falcaro, P. Using functional nano- and microparticles for the preparation of metal-organic framework composites with novel properties. *Accounts Chem. Res.* **2014**, *47*, 396–405. [CrossRef] [PubMed]

20. Hu, M.; Torad, N.L.K.; Chiang, Y.-D.; Wu, K.C.W.; Yamauchi, Y. Size- and shape-controlled synthesis of Prussian Blue nanoparticles by a polyvinylpyrrolidone-assisted crystallization process. *CrystEngComm* **2012**, *14*, 3387–3396.

21. Fan, L.; Chen, H.; Xiao, D.; Wang, E. Synthesis, structure, and characterization of a new metal-organic framework containing meso-helices. *Z. Anorg. Allg. Chem.* **2013**, *639*, 558–562. [CrossRef]

22. Zhang, W.; Zhao, Y.; Malgras, V.; Ariga, K.; Yamauchi, Y.; Liu, J.; Jiang, J.S.; Hu, M. Synthesis of monocrystalline nanoframes of prussian blue analogues by controlled preferential etching. *Angew. Chem. Int. Ed.* **2016**, *55*, 8228–8234. [CrossRef] [PubMed]

23. Ariga, K.; Malgras, V.; Ji, Q.; Zakaria, M.B.; Yamauchi, Y. Coordination nanoarchitectonics at interfaces between supramolecular and materials chemistry. *Coordin. Chem. Rev.* **2016**, *320*, 139–152. [CrossRef]

24. Felts, A.C.; Andrus, M.J.; Knowles, E.S.; Quintero, P.K.; Ahir, A.R.; Risset, O.N.; Li, C.H.; Maurin, I.; Halder, G.J.; Abboud, K.A.; et al. Evidence for interface-induced strain and its influence on photomagnetism in prussian blue analogue core-shell heterostructures, RbaCob Fe(CN)(6) (c)center dot mH(2)O@KjNik Cr(CN)(6) (l)center dot nH(2)O. *J. Phys. Chem. C* **2016**, *120*, 5420–5429. [CrossRef]

25. Yue, Y.; Zhang, Z.; Binder, A.J.; Chen, J.; Jin, X.; Overbury, S.H.; Dai, S. Hierarchically superstructured prussian blue analogues: Spontaneous assembly synthesis and applications as pseudocapacitive materials. *Chem. Sus. Chem.* **2015**, *8*, 177–183. [CrossRef] [PubMed]

26. Kim, D.S.; Zakaria, M.B.; Park, M.-S.; Alowasheeir, A.; Alshehri, S.M.; Yamauchi, Y.; Kim, H. Dual-textured Prussian Blue nanocubes as sodium ion storage materials. *Electrochim. Acta* **2017**, *240*, 300–306. [CrossRef]

27. Yue, Y.; Binder, A.J.; Guo, B.; Zhang, Z.; Qiao, Z.A.; Tian, C.; Dai, S. Mesoporous Prussian blue analogues: Template-free synthesis and sodium-ion battery applications. *Angew. Chem. Int. Ed.* **2014**, *53*, 3134–3137. [CrossRef] [PubMed]

28. Lu, X.; Au, L.; McLellan, J.; Li, Z.-Y.; Marquez, M.; Xia, Y. Fabrication of cubic nanocages and nanoframes by dealloying Au/Ag alloy nanoboxes with an aqueous etchant based on Fe(NO$_3$)(3) or NH$_4$OH. *Nano lett.* **2007**, *7*, 1764–1769. [CrossRef] [PubMed]

29. Mahmoud, M.A.; El-Sayed, M.A. Aggregation of gold nanoframes reduces, rather than enhances, SERS efficiency due to the trade-off of the inter- and intraparticle plasmonic fields. *Nano lett.* **2009**, *9*, 3025–3031. [CrossRef] [PubMed]

30. Mahmoud, M.A.; El-Sayed, M.A. Gold nanoframes: Very high surface plasmon fields and excellent near-infrared sensors. *J. Am. Chem. Soc.* **2010**, *132*, 12704–12710. [CrossRef] [PubMed]

31. Sui, Y.; Fu, W.; Zeng, Y.; Yang, H.; Zhang, Y.; Chen, H.; Li, Y.; Li, M.; Zou, G. Synthesis of Cu$_2$O nanoframes and nanocages by selective oxidative etching at room temperature. *Angew. Chem. Int. Ed.* **2010**, *49*, 4282–4285. [CrossRef] [PubMed]

32. McEachran, M.; Keogh, D.; Pietrobon, B.; Cathcart, N.; Gourevich, I.; Coombs, N.; Kitaev, V. Ultrathin gold nanoframes through surfactant-free templating of faceted pentagonal silver nanoparticles. *J. Am. Chem. Soc.* **2011**, *133*, 8066–8069. [CrossRef] [PubMed]

33. Hong, X.; Wang, D.; Cai, S.; Rong, H.; Li, Y. Single-crystalline octahedral au-ag nanoframes. *J. Am. Chem. Soc.* **2012**, *134*, 18165–18168. [CrossRef] [PubMed]

34. Xie, S.; Lu, N.; Xie, Z.; Wang, J.; Kim, M.J.; Xia, Y. Synthesis of Pd-Rh core-frame concave nanocubes and their conversion to rh cubic nanoframes by selective etching of the Pd cores. *Angew. Chem. Int. Ed.* **2012**, *51*, 10266–10270. [CrossRef] [PubMed]

35. Chen, C.; Kang, Y.; Huo, Z.; Zhu, Z.; Huang, W.; Xin, H.L.; Snyder, J.D.; Li, D.; Herron, J.A.; Mavrikakis, M.; et al. Highly crystalline multimetallic nanoframes with three-dimensional electrocatalytic surfaces. *Science* **2014**, *343*, 1339–1343. [CrossRef] [PubMed]

36. Zhang, L.; Roling, L.T.; Wang, X.; Vara, M.; Chi, M.; Liu, J.; Choi, S.-I.; Park, J.; Herron, J.A.; Xie, Z.; et al. Platinum-based nanocages with subnanometer-thick walls and well-defined, controllable facets. *Science* **2015**, *349*, 412–416. [CrossRef] [PubMed]

37. Wessells, C.D.; McDowell, M.T.; Peddada, S.V.; Pasta, M.; Huggins, R.A.; Cui, Y. Tunable reaction potentials in open framework nanoparticle battery electrodes for grid-scale energy storage. *Acs Nano* **2012**, *6*, 1688–1694. [CrossRef] [PubMed]

38. Pasta, M.; Wessells, C.D.; Liu, N.; Nelson, J.; McDowell, M.T.; Huggins, R.A.; Toney, M.F.; Cui, Y. Full open-framework batteries for stationary energy storage. *Nat. Commun.* **2014**, *5*, 10. [CrossRef] [PubMed]

39. Wang, R.Y.; Wessells, C.D.; Huggins, R.A.; Cui, Y. Highly reversible open framework nanoscale electrodes for divalent ion batteries. *Nano lett.* **2013**, *13*, 5748–5752. [CrossRef] [PubMed]

40. Lee, H.-W.; Wang, R.Y.; Pasta, M.; Lee, S.W.; Liu, N.; Cui, Y. Manganese hexacyanomanganate open framework as a high-capacity positive electrode material for sodium-ion batteries. *Nat. Commun.* **2014**, *5*. [CrossRef] [PubMed]

41. Chae, M.S.; Hyoung, J.; Jang, M.; Lee, H.; Hong, S.-T. Potassium nickel hexacyanoferrate as a high-voltage cathode material for nonaqueous magnesium-ion batteries. *J. Power Sources* **2017**, *363*, 269–276. [CrossRef]

42. Okubo, M.; Li, C.H.; Talham, D.R. High rate sodium ion insertion into core-shell nanoparticles of Prussian blue analogues. *Chem. Commun.* **2014**, *50*, 1353–1355. [CrossRef] [PubMed]

43. Tang, W.; Zhu, Y.; Hou, Y.; Liu, L.; Wu, Y.; Loh, K.P.; Zhang, H.; Zhu, K. Aqueous rechargeable lithium batteries as an energy storage system of superfast charging. *Energ. Environ. Sci.* **2013**, *6*, 2093–2104. [CrossRef]

44. Zhao, F.; Wang, Y.; Xu, X.; Liu, Y.; Song, R.; Lu, G.; Li, Y. Cobalt hexacyanoferrate nanoparticles as a high-rate and ultra-stable supercapacitor electrode material. *ACS Appl. Mater. Inter.* **2014**, *6*, 11007–11012. [CrossRef] [PubMed]

45. Pasta, M.; Wessells, C.D.; Huggins, R.A.; Cui, Y. A high-rate and long cycle life aqueous electrolyte battery for grid-scale energy storage. *Nat. Commun.* **2012**, *3*, 1149–1155. [CrossRef] [PubMed]

46. Wessells, C.D.; Huggins, R.A.; Cui, Y. Copper hexacyanoferrate battery electrodes with long cycle life and high power. *Nat. Commun.* **2011**, *2*, 550–554. [CrossRef] [PubMed]

47. Chen, R.; Huang, Y.; Xie, M.; Zhang, Q.; Zhang, X.; Li, L.; Wu, F. Preparation of Prussian Blue submicron particles with a pore structure by two-step optimization for Na-ion battery cathodes. *ACS Appl. Mater. Inter.* **2016**, *8*, 16078–16086. [CrossRef] [PubMed]
48. Wang, L.; Lu, Y.; Liu, J.; Xu, M.; Cheng, J.; Zhang, D.; Goodenough, J.B. A superior low-cost cathode for a Na-ion battery. *Angew. Chem. Int. Ed.* **2013**, *52*, 1964–1967. [CrossRef] [PubMed]
49. You, Y.; Wu, X.-L.; Yin, Y.-X.; Guo, Y.-G. High-quality Prussian blue crystals as superior cathode materials for room-temperature sodium-ion batteries. *Energ. Environ. Sci.* **2014**, *7*, 1643–1647. [CrossRef]
50. Liu, Y.; Qiao, Y.; Zhang, W.; Li, Z.; Ji, X.; Miao, L.; Yuan, L.; Hu, X.; Huang, Y. Sodium storage in Na-rich NaxFeFe (CN)(6) nanocubes. *Nano Energy* **2015**, *12*, 386–393. [CrossRef]
51. You, Y.; Yu, X.; Yin, Y.; Nam, K.-W.; Guo, Y.-G. Sodium iron hexacyanoferrate with high Na content as a Na-rich cathode material for Na-ion batteries. *Nano Res.* **2014**, *8*, 117–128. [CrossRef]
52. Wessells, C.D.; Peddada, S.V.; Huggins, R.A.; Cui, Y. Nickel hexacyanoferrate nanoparticle electrodes for aqueous sodium and potassium ion batteries. *Nano Lett.* **2011**, *11*, 5421–5425. [CrossRef] [PubMed]
53. Targholi, E.; Mousavi-Khoshdel, S.M.; Rahmanifara, M.; Yahya, M.Z.A. Cu- and Fe-hexacyanoferrate as cathode materials for Potassium ion battery: A First-principles study. *Chem. Phys. Lett.* **2017**, *687*, 244–249. [CrossRef]
54. Lu, Y.; Wang, L.; Cheng, J.; Goodenough, J.B. Prussian blue: A new framework of electrode materials for sodium batteries. *Chem. Commun.* **2012**, *48*, 6544–6546. [CrossRef] [PubMed]
55. Wessells, C.D.; Peddada, S.V.; McDowell, M.T.; Huggins, R.A.; Cui, Y. The effect of insertion species on nanostructured open framework hexacyanoferrate battery electrodes. *J. Electrochem. Soc.* **2012**, *159*, 98–103. [CrossRef]
56. Kim, D.; Hwang, T.; Lim, J.-M.; Park, M.-S.; Cho, M.; Cho, K. Hexacyanometallates for sodium-ion batteries: Insights into higher redox potentials using d electronic spin configurations. *Phys. Chem. Chem.l Phys.* **2017**, *19*, 10443–10452. [CrossRef] [PubMed]
57. Liu, Y.; He, D.; Han, R.; Wei, G.; Qiao, Y. Nanostructured potassium and sodium ion incorporated Prussian blue frameworks as cathode materials for sodium-ion batteries. *Chem. Commun.* **2017**, *53*, 5569–5572. [CrossRef] [PubMed]
58. Liang, G.; Xu, J.; Wang, X. Synthesis and characterization of organometallic coordination polymer nanoshells of Prussian blue using miniemulsion periphery polymerization (MEPP). *J. Am. Chem. Soc.* **2009**, *131*, 5378–5379. [CrossRef] [PubMed]
59. McHale, R.; Ghasdian, N.; Liu, Y.; Ward, M.B.; Hondow, N.S.; Wang, H.; Miao, Y.; Brydson, R.; Wang, X. Prussian blue coordination polymer nanobox synthesis using miniemulsion periphery polymerization (MEPP). *Chem. Commun.* **2010**, *46*, 4574–4576. [CrossRef] [PubMed]
60. Hu, M.; Jiang, J.S. Facile synthesis of air-stable Prussian white microcubes via a hydrothermal method. *Mater. Res. Bull.* **2011**, *46*, 702–707. [CrossRef]
61. McHale, R.; Liu, Y.; Ghasdian, N.; Hondow, N.S.; Ye, S.; Lu, Y.; Brydson, R.; Wang, X. Dual lanthanide role in the designed synthesis of hollow metal coordination (Prussian Blue analogue) nanocages with large internal cavity and mesoporous cage. *Nanoscale* **2011**, *3*, 3685–3694. [CrossRef] [PubMed]
62. Roy, X.; Hui, J.K.H.; Rabnawaz, M.; Liu, G.; MacLachlan, M.J. Prussian blue nanocontainers: selectively permeable hollow metal-organic capsules from block Ionomer emulsion-induced assembly. *J. Am. Chem. Soc.* **2011**, *133*, 8420–8423. [CrossRef] [PubMed]
63. Risset, O.N.; Knowles, E.S.; Ma, S.; Meisel, M.W.; Talham, D.R. RbjMk Fe(CN)(6) (l) (M = Co, Ni) Prussian blue analogue hollow nanocubes: A new example of a multilevel pore system. *Chem. Mater.* **2013**, *25*, 42–47. [CrossRef]
64. Shen, Q.; Jiang, J.; Fan, M.; Liu, S.; Wang, L.; Fan, Q.; Huang, W. Prussian blue hollow nanostructures: Sacrificial template synthesis and application in hydrogen peroxide sensing. *J. Electroanal. Chem.* **2014**, *712*, 132–138. [CrossRef]
65. Hu, M.; Jiang, J.-S.; Zeng, Y. Prussian blue microcrystals prepared by selective etching and their conversion to mesoporous magnetic iron(III) oxides. *Chem. Commun.* **2010**, *46*, 1133–1135. [CrossRef] [PubMed]
66. Hu, M.; Torad, N.L.; Yamauchi, Y. Preparation of Various Prussian Blue analogue hollow nanocubes with single crystalline shells. *Eur. J. Inorg. Chem.* **2012**, *30*, 4795–4799. [CrossRef]
67. Lee, S.-H.; Huh, Y.-D. Preferential evolution of Prussian blue's morphology from cube to hexapod. *B. Kor. Chem. Soc.* **2012**, *33*, 1078–1080. [CrossRef]

68. Song, Y.; He, J.; Wu, H.; Li, X.; Yu, J.; Zhang, Y.; Wang, L. Preparation of porous hollow CoOx nanocubes via chemical etching prussian blue analogue for glucose sensing. *Electrochim. Acta* **2015**, *182*, 165–172. [CrossRef]
69. Han, L.; Yu, T.; Lei, W.; Liu, W.; Feng, K.; Ding, Y.; Jiang, G.; Xu, P.; Chen, Z. Nitrogen-doped carbon nanocones encapsulating with nickel-cobalt mixed phosphides for enhanced hydrogen evolution reaction. *J. Mater. Chem. A* **2017**, *5*, 16568–16572. [CrossRef]
70. Liu, Y.; Wei, G.; Ma, M.; Qiao, Y. Role of acid in tailoring Prussian blue as cathode for high-performance sodium-ion battery. *Chem. Eur. J.* **2017**, *23*, 15991–15996. [CrossRef] [PubMed]
71. Zhang, G.; Xu, X.; Ji, Q.; Liu, R.; Liu, H.; Qiu, J.; Li, J. Porous nanobimetallic Fe-Mn cubes with high valent mn and highly efficient removal of arsenic(III). *ACS Appl. Mater. Inter.* **2017**, *9*, 14868–14877. [CrossRef] [PubMed]
72. Hu, M.; Furukawa, S.; Ohtani, R.; Sukegawa, H.; Nemoto, Y.; Reboul, J.; Kitagawa, S.; Yamauchi, Y. Synthesis of Prussian blue nanoparticles with a hollow interior by controlled chemical etching. *Angew. Chem. Int. Ed.* **2012**, *51*, 984–988. [CrossRef] [PubMed]
73. Yang, J.; Zhang, F.; Lu, H.; Hong, X.; Jiang, H.; Wu, Y.; Li, Y. Hollow Zn/Co ZIF particles derived from core-shell zif-67@zif-8 as selective catalyst for the semi-hydrogenation of acetylene. *Angew. Chem. Int. Ed.* **2015**, *54*, 10889–10893. [CrossRef] [PubMed]
74. Ren, W.; Qin, M.; Zhu, Z.; Yan, M.; Li, Q.; Zhang, L.; Liu, D.; Mai, L. Activation of sodium storage sites in prussian blue analogues via surface etching. *Nano lett.* **2017**, *17*, 4713–4718. [CrossRef] [PubMed]
75. Han, L.; Yu, X.-Y.; Lou, X.W. Formation of Prussian-blue-analog nanocages via a direct etching method and their conversion into Ni-Co-Mixed oxide for enhanced oxygen evolution. *Adv. Mater.* **2016**, *28*, 4601–4605. [CrossRef] [PubMed]
76. Ahirwar, P.; Clark, S.P.R.; Patel, V.; Rotter, T.J.; Hains, C.; Albrecht, A.; Dawson, L.R.; Balakrishnan, G. Perforated (In)GaSb quantum wells on GaSb substrates through the use of As(2) based in situ etches. *J. Vac. Sci. Technol. B* **2011**, *29*. [CrossRef]
77. Li, G.; Li, Y.; Li, Y.; Peng, H.; Chen, K. Polyaniline nanorings and flat hollow capsules synthesized by in situ sacrificial oxidative templates. *Macromolecules* **2011**, *44*, 9319–9323. [CrossRef]
78. Yang, K.C.; Jeon, M.H.; Yeom, G.Y. A study on the etching characteristics of magnetic tunneling junction materials using DC pulse-biased inductively coupled plasmas. *Jpn. J. Appl. Phys.* **2015**, *54*. [CrossRef]
79. Zhang, C.; Zhou, Y.; Zhang, Y.; Zhao, S.; Fang, J.; Sheng, X.; Zhang, H. A novel hierarchical TiO2@Pt@mSiO(2) hollow nanocatalyst with enhanced thermal stability. *J. Alloy. Compd.* **2017**, *701*, 780–787. [CrossRef]
80. Chen, L.; Bao, J.L.; Dong, X.; Truhlar, D.G.; Wang, Y.; Wang, C.; Xia, Y. Aqueous Mg-Ion battery based on polyimide anode and Prussian blue cathode. *ACS Energy Lett.* **2017**, *2*, 1115–1121. [CrossRef]
81. Wu, X.; Sun, M.; Guo, S.; Qian, J.; Liu, Y.; Cao, Y.; Ai, X.; Yang, H. Vacancy-free Prussian blue nanocrystals with high capacity and superior cyclability for aqueous sodium-ion batteries. *Chem. Nano Mat.* **2015**, *1*, 188–193. [CrossRef]
82. Whitacre, J.F.; Tevar, A.; Sharma, S. Na4Mn9O18 as a positive electrode material for an aqueous electrolyte sodium-ion energy storage device. *Electrochem. Commun.* **2010**, *12*, 463–466. [CrossRef]
83. Wu, X.; Cao, Y.; Ai, X.; Qian, J.; Yang, H. A low-cost and environmentally benign aqueous rechargeable sodium-ion battery based on NaTi2(PO4)3–Na2NiFe(CN)6 intercalation chemistry. *Electrochem. Commun.* **2013**, *31*, 145–148. [CrossRef]
84. Luo, J.-Y.; Cui, W.-J.; He, P.; Xia, Y.-Y. Raising the cycling stability of aqueous lithium-ion batteries by eliminating oxygen in the electrolyte. *Nat. Chem.* **2010**, *2*, 760–765. [CrossRef] [PubMed]
85. Hu, M.; Ishihara, S.; Ariga, K.; Imura, M.; Yamauchi, Y. Kinetically controlled crystallization for synthesis of monodispersed coordination polymer nanocubes and their self-assembly to periodic arrangements. *Chem. Eur. J.* **2013**, *19*, 1882–1885. [CrossRef] [PubMed]
86. Widmann, A.; Kahlert, H.; Wulff, H.; Scholz, F. Electrochemical and mechanochemical formation of solid solutions of potassium copper(II)/zinc(II) hexacyanocobaltate(III)/hexacyanoferrate(III) KCuxZn1-x hcc (x) hcf (1-x). *J. Solid State Electr.* **2005**, *9*, 380–389. [CrossRef]
87. Kulesza, P.J.; Malik, M.A.; Denca, A.; Strojek, J. In situ FT-IR/ATR spectroelectrochemistry of Prussian blue in the solid state. *Anal. Chem.* **1996**, *68*, 2442–2446. [CrossRef]
88. Hung, T.-F.; Chou, H.-L.; Yeh, Y.-W.; Chang, W.-S.; Yang, C.-C. Combined experimental and computational studies of a na2ni1-xcuxfe(cn)(6) cathode with tunable potential for aqueous rechargeable sodium-ion batteries. *Chem. Europe. J.* **2015**, *21*, 15686–15691. [CrossRef] [PubMed]

89. Hung, T.-F.; Cheng, W.-J.; Chang, W.-S.; Yang, C.-C.; Shen, C.-C.; Kuo, Y.-L. Ascorbic acid-assisted synthesis of mesoporous sodium vanadium phosphate nanoparticles with highly sp(2)-coordinated carbon coatings as efficient cathode materials for rechargeable sodium-ion batteries. *Chem. Eur. J.* **2016**, *22*, 10620–10626. [CrossRef] [PubMed]

90. You, Y.; Wu, X.-L.; Yin, Y.-X.; Guo, Y.-G. A zero-strain insertion cathode material of nickel ferricyanide for sodium-ion batteries. *J. Mater. Chem. A* **2013**, *1*, 14061–14065. [CrossRef]

crystals

MDPI

Article

Incorporating the Thiazolo[5,4-d]thiazole Unit into a Coordination Polymer with Interdigitated Structure

Simon Millan, Gamall Makhloufi and Christoph Janiak *

Institut für Anorganische Chemie und Strukturchemie, Heinrich-Heine-Universität, Universitätsstraße 1, 40225 Düsseldorf, Germany; simon.millan@hhu.de (S.M.); gamall.makhloufi@hhu.de (G.M.)
* Correspondence: janiak@hhu.de; Tel.: +49-211-81-12286

Received: 21 December 2017; Accepted: 6 January 2018; Published: 12 January 2018

Abstract: The linker 2,5-di(4-pyridyl)thiazolo[5,4-d]thiazole (Dptztz), whose synthesis and structure is described here, was utilized together with benzene-1,3-dicarboxylate (isophthalate, $1,3\text{-BDC}^{2-}$) for the preparation of the two-dimensional coordination network [Zn(1,3-BDC)Dptztz]·DMF (DMF = dimethylformamide) via a solvothermal reaction. Compound [Zn(1,3-BDC)Dptztz]·DMF belongs to the class of coordination polymers with interdigitated structure (CIDs). The incorporated DMF solvent molecules can be removed through solvent exchange and evacuation such that the supramolecular 3D packing of the 2D networks retains porosity for CO_2 adsorption in activated [Zn(1,3-BDC)Dptztz]. The first sorption study of a tztz-functionalized porous metal-organic framework material yields a BET surface of 417 m^2/g calculated from the CO_2 adsorption data. The heat of adsorption for CO_2 exhibits a relative maximum with 27.7 kJ/mol at an adsorbed CO_2 amount of about 4 cm^3/g STP, which is interpreted as a gate-opening effect.

Keywords: coordination polymer; MOF; gate-opening; thiazolo[5,4-d]thiazole; mixed-ligand

---

1. Introduction

Metal-organic frameworks are an intensively studied class of porous materials. Due to the immense quantity of possible inorganic and organic building units several applications are discussed (e.g., gas storage and separation, catalysis, sensing and heat transformation) [1–13]. Many different organic functionalities have been introduced into the frameworks either by a priori ligand functionalization or by post synthetic modification [14–16]. 4,4′-Biypyridine based ligands have been used to construct a diverse set of different topologies (e.g., one-dimensional chains, ladders, two-dimensional and three-dimensional networks) [17]. 4,4′-Bipyridine ligands are also widely used in the synthesis of open network structures in combination with dicarboxylate ligands (e.g., terephthalate, isophthalate) via the so called mixed-ligand strategy [18,19]. Through functionalization of the organic ligands, the pore surface of those mixed-ligand MOFs can be tuned to enhance the selectivity in their sorption or sensing properties [20–23]. One famous family of mixed-ligands MOFs are the CIDs (coordination polymers with interdigitated structure) popularized by Kitagawa and co-workers. CIDs consist of an angular ligand (e.g., isophthalate, benzophenone-4,4′-dicarboxylate, azulene-1,6-dicarboxylate) and a 4,4′-bipyridine derivative and divalent transition metal ions. CIDs show very intriguing sorption properties due to their potential for functionalization and often inherent structural flexibility [24–29].

The heterocyclic thiazolo[5,4-d]thiazole (tztz) system (Figure 1) experienced a renaissance in the last decade after it was first prepared by Ketcham et al. in the 1960s [30]. The tztz unit was incorporated into photoactive materials due to its rigid and planar structure and electron deficiency. Both Maes et al. and Dessi et al. reviewed the synthetic procedures to obtain tztz-containing small molecules and polymers as well as their application in the field of organic electronics (e.g., OFETs, OSCs) [31,32]. In contrast, the tztz unit has been reported only in relatively few coordination compounds (15 hits in the CCDC database). The first examples were ruthenium and copper complexes with the doubly chelating 2,5-di(2-pyridyl)thiazolo[5,4-d]thiazole synthesized by Steel et al. [33]. Coordination polymers with 2,5-thiazolo[5,4-d]thiazoledicarboxylic acid (Figure 1) were obtained by Cheetham et al. with alkaline earth metals, whose connectivities vary with the cation size, and by Palmisano et al. with some transition metals, in which the ligand shows a *N,O*-chelating mode [34,35]. D'Alessandro et al. incorporated the donor-acceptor ligand *N,N'*-(thiazolo[5,4-d]thiazole-2,5-diylbis(4,1-phenylene))bis(*N*-(pyridine-4-yl)pyridin-4-amine into a two-dimensional zinc MOF and studied its electrochemical properties [36]. Recently the same group published the spectroelectrochemical properties of a ruthenium coordination complex with this ligand [37]. Additionally, Dai et al. synthesized tztz-linked microporous organic polymers, which show a high CO_2:N_2 selectivity [38].

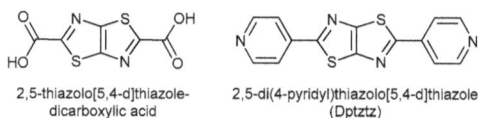

2,5-thiazolo[5,4-d]thiazole-
dicarboxylic acid

2,5-di(4-pyridyl)thiazolo[5,4-d]thiazole
(Dptztz)

Figure 1. Examples of tztz-containing ligands.

Herein, we present the synthesis, structural analysis and the sorption properties of a new coordination polymer with interdigitated structure of the formula [Zn(1,3-BDC)Dptztz] consisting of Zn^{2+} ions, isophthalate and the 4,4'-bipyridine derivative 2,5-di(4-pyridyl)thiazolo[5,4-d]thiazole (Dptztz) (Figure 1, Scheme 1).

Scheme 1. Reaction scheme for the synthesis of 2,5-di(4-pyridyl)thiazolo[5,4-d]thiazole from 4-pyridinecarboxaldehyde and dithiooxamide.

2. Materials and Methods

The chemicals used were obtained from commercial sources. No further purification has been carried out. CHN analysis was performed with a Perkin Elmer CHN 2400 (Perkin Elmer, Waltham, MA, USA). IR-spectra were recorded on a Bruker Tensor 37 IR spectrometer (Bruker Optics, Ettlingen, Germany) with ATR unit. Thermogravimetric analysis (TGA) was done with a Netzsch TG 209 F3 Tarsus (Netzsch, Selb, Germany) in the range from 20 to 700 °C, equipped with Al-crucible and applying a heating rate of 10 K·min^{-1} under nitrogen. The melting point was determined using a Büchi Melting Point apparatus B540. The powder X-ray diffraction pattern (PXRD) was obtained on a Bruker D2 Phaser powder diffractometer with a flat silicon, low background sample holder, at 30 kV, 10 mA for Cu-Kα radiation (λ = 1.5418 Å). Sorption isotherms were measured using a Micromeritics ASAP 2020 automatic gas sorption analyzer equipped with oil-free vacuum pumps (ultimate vacuum <10^{-8} mbar) and valves, which guaranteed contamination free measurements. The sample was connected to the preparation port of the sorption analyzer and degassed under vacuum until the outgassing rate, i.e., the rate of pressure rise in the temporarily closed manifold with the connected sample tube, was less than 2 µTorr/min at the specified temperature of 120 °C. After weighing, the sample tube was then transferred to the analysis port of the sorption analyzer. All used gases (He, N_2, CO_2) were of ultra-high purity (UHP, grade 5.0, 99.999%) and the STP volumes are given according to the NIST standards (293.15 K, 101.325 kPa). Helium gas was used for the determination of the cold and warm free space of the sample tubes. The heat of adsorption values were calculated using the ASAP

2020 v3.05 software. Water sorption isotherms were obtained volumetrically from a Quantachrome Autosorb iQ MP instrument equipped with an all-gas option. Prior to the sorption experiments, the compounds were degassed under dynamic vacuum.

2,5-Di(4-pyridyl)thiazolo[5,4-d]thiazole (Dptztz): 1.02 g (8.5 mmol) dithiooxamide and 2.0 mL (22 mmol, 2.6 eq) 4-pyridinecarboxaldehyde in 50 mL anhydrous DMF were refluxed for 6.5 h under nitrogen. During cooling the reaction mixture to room temperature, the product crystallized out in form of yellow prisms. The ligand was filtered and washed with a small amount of DMF and extensively with water. After drying in a vacuum oven at 60 °C overnight 1.82 g (6.1 mmol, 72 %) were obtained. ^{1}H-NMR (300 MHz, DMSO-d$_6$) δ [ppm]: 8.78 (d, $^4J_{H,H}$ = 6.12 Hz, 2H), 7.88 (d, $^4J_{H,H}$ = 6.12 Hz, 2H); MS (EI) m/z [rel. int.]: 296 (100%); 87.9 (91%); mp 319–322 °C.

[Zn(1,3-BDC)Dptztz]: 5.0 mg (0.017 mmol) of Dptztz were dissolved in 3 mL of hot DMF in a Pyrex tube. 5.4 mg (0.020 mmol) of Zn(NO$_3$)$_2$·4H$_2$O and 2.8 mg (0.020 mmol) of isophthalic acid dissolved in 2 mL of DMF were added to the warm solution. The Pyrex tube was capped and placed into a preheated isothermal oven at 80 °C. After 12 h the first crystals appeared. After 3 days, the sample was removed from the oven and the solvent was directly exchanged with 3 × 3 mL of hot DMF. A light yellow crystal was selected to collect the single crystal data. Yield: 4 mg.

A larger amount of material was prepared by dissolving 100.4 mg (0.34 mmol) of Dptztz in 40 mL of hot DMF in a 100 mL Schott vial. Afterwards 88.4 mg (0.34 mmol) of Zn(NO$_3$)·4H$_2$O and 56.6 mg (0.34 mmol) of isophthalic acid dissolved in 10 mL of DMF were added and placed in an isothermal oven preheated at 120 °C. The sample was taken out after 3 days and the solvent was directly exchanged with 3 × 20 mL of hot DMF. Yield: 182.6 mg (90%). EA [%] calc. for: C$_{22}$H$_{12}$N$_4$O$_4$S$_2$Zn C 50.25, H 2.30 N 10.70; found: C 50.89, H 2.93, N 11.51. IR (ATR) $\tilde{v}_{max}$ [cm^{-1}]: 3433, 1608, 1564, 1443, 1391, 1213, 1014, 833, 743, 724, 662, 619, 510.

Single Crystal X-ray Structures

Suitable crystals were carefully selected under a polarizing microscope, covered in protective oil and mounted on a 0.05 mm cryo loop. *Data collection*: Bruker Kappa APEX2 CCD X-ray diffractometer (Bruker AXS Inc., Madison, WI, USA) with microfocus tube, Mo-Kα radiation (λ = 0.71073 Å), multi-layer mirror system, ω-scans; data collection with APEX2 [39], cell refinement with SMART and data reduction with SAINT [39], experimental absorption correction with SADABS [40]. *Structure analysis and refinement*: All structures were solved by direct methods using SHELXL2014 [41,42]; refinement was done by full-matrix least squares on F^2 using the SHELX-97 program suite. The hydrogen atoms for aromatic CH and for the amide group in DMF were positioned geometrically (C-H = 0.95 Å) and refined using a riding model (AFIX 43) with U$_{iso}$(H) = 1.2U$_{eq}$(C). The hydrogen atoms for CH$_3$ of DMF were positioned geometrically (C-H = 0.98 Å) and refined using a riding model (AFIX 137) with U$_{iso}$(H) = 1.5U$_{eq}$(C). In [Zn(1,3-BDC)Dptztz] the thiazolothiazol (tztz) moiety was refined with a disorder model corresponding to a ring flip, which exchanges the S and N orientation, using PART n commands. The occupation factors of the S and N atoms were refined to about 0.904 for the A atoms and 0.096 for the B atoms. Thus, the disorder is relatively minor with only about 9.6% of the S and N atoms in the flipped position. The major atom tztz positions are designated as "A", the minor ones as "B". The minor positions could only be refined isotropically. The DMF crystal solvent molecule contained disorder in the two methyl groups, with two positions for each methyl group. This disorder does not give a perfect oriented Me$_2$N group but we decided to keep the slightly disordered DMF molecule instead of removing its contribution with SQUEEZE. Each methyl group disorder was refined independently concerning the occupation factors. Crystal data and details on the structure refinement are given in Table 1. Graphics were drawn with DIAMOND [43]. Analyses on the supramolecular interaction were done with PLATON [44]. Phase purity and the representative nature of the single crystal was verified by positively matching the simulated and experimental powder X-ray diffractogram (PXRD) of the as-synthesized sample (Figure 2). CCDC 1812892 and 1812893 contain the supplementary crystallographic data for this paper. These data can be obtained free of charge via http://www.ccdc.cam.ac.uk/conts/retrieving.html.

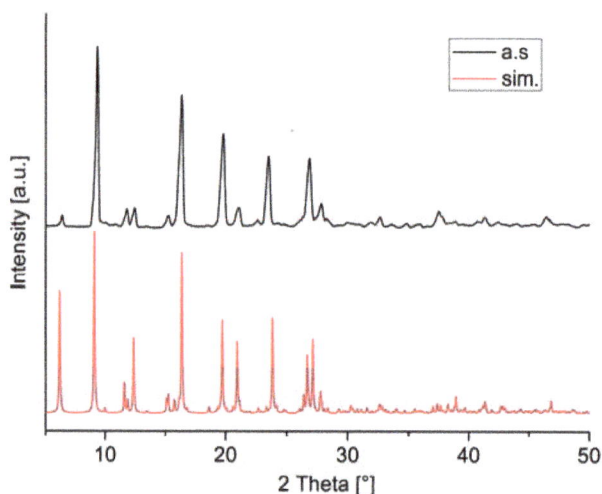

Figure 2. PXRD pattern of [Zn(1,3-BDC)Dptztz]·DMF (simulated (**red**), as-synthesized (**black**)).

Table 1. Crystal data and refinement details.

	Dptztz	[Zn(1,3 BDC)Dptztz]·DMF
Chemical formula	$C_{14}H_8N_4S_2$	$C_{22}H_{12}N_4O_4S_2Zn\cdot C_3H_7NO$
Mr	296.36	598.94
Crystal system, space group	Monoclinic, $P2_1/c$	Triclinic, $P\bar{1}$
Temperature (K)	100	100
a (Å)	8.3873 (5)	9.1388 (6)
b (Å)	6.3140 (3)	10.0354 (7)
c (Å)	11.7170 (6)	14.2804 (11)
α (°)	90	88.417 (4)
β (°)	93.699 (3)	88.236 (5)
γ (°)	90	75.636 (4)
V (Å^3)	619.21 (6)	1267.86 (16)
Z	2	2
μ (mm^{-1})	0.423	1.181
Crystal size (mm)	$0.10 \times 0.05 \times 0.05$	$0.10 \times 0.05 \times 0.01$
Absorption correction	Multi-scan, wR2(int) was 0.1649 before and 0.0771 after correction. The Ratio of minimum to maximum transmission is 0.8473. The $\lambda/2$ correction factor is 0.0015.	Multi-scan, wR2(int) was 0.1533 before and 0.0488 after correction. The Ratio of minimum to maximum transmission is 0.9318. The $\lambda/2$ correction factor is 0.0015.
T_{min}, T_{max}	0.6330, 0.7471	0.6951, 0.7460
No. of measured, independent and observed reflections	6837, 965, 847 [I > 2σ(I)]	17151, 4743, 3696 [I > 2σ(I)]
R_{int}	0.049	0.045
(sin θ/λ)max (Å^{-1})	0.639	0.612
R, wR(F^2), S [F^2 > 2σ (F^2)] R, wR(F^2), S [all data]	0.0284, 0.0675, 1.067 0.359, 0.0699, 1.067	0.0400, 0.0849, 1.055 0.0609, 0.0916, 1.055
No. of reflections	965	4743
No. of parameters	91	364
$\Delta\varrho_{max}$, $\Delta\varrho_{min}$ (e·Å^{-3})	0.238, -0.182	0.645, −0.581

3. Results and Discussion

2,5-Di(4-pyridyl)thiazolo[5,4-d]thiazole (Dptztz) was synthesized according to the literature by the condensation of 4-pyridinecarboxaldehyde and dithioaxamide (Scheme 1) [45,46].

Single crystals of Dptztz were obtained after recrystallization from DMF in form of yellow prisms. Dptztz crystallizes in the monoclinic space group $P2_1/c$ with half of the molecule in the asymmetric unit as the molecule sits on an inversion center (Figure 3). The molecule is almost planar with a dihedral angle between the pyridine ring and the tztz moiety of $13.65°$.

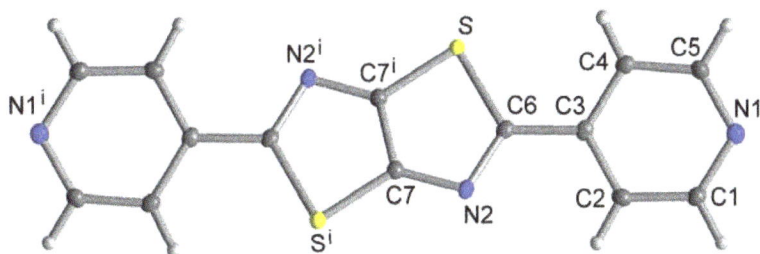

Figure 3. Molecular structure of Dptztz (50% thermal ellipsoids, symmetry transformation $1 - x$, $2 - y$, $2 - z$).

Complementary CH$\cdots$N hydrogen bonds between N1 and C1-H1 of adjacent Dptztz molecules form 1D strands which are parallel displaced by π-π interactions (Figure S1 in Supplementary Material).

Single crystals of the coordination network [Zn(1,3-BDC)Dptztz]·DMF were obtained after three days from a solvothermal reaction of Zn(NO$_3$)$_2$·4H$_2$O, isophthalic acid and Dptztz in a molar ratio 1:1:1 in DMF at 80 °C. Due to the low solubility of Dptztz in common organic solvents, the reaction was carried out in a concentration of 3.4×10^{-3} mol/L and the mother liquor was directly exchanged with hot DMF after the crystallization process to remove unreacted Dptztz ligand. A larger amount of material for the sorption experiments was synthesized by scaling up the reaction by the factor of twenty in twice the concentration (6.8×10^{-3} mol/L) at 120 °C.

The crystal structure of the two-dimensional (2D) coordination network [Zn(1,3-BDC)Dptztz]·DMF was determined by single crystal diffraction analysis at 100 K. Compound [Zn(1,3-BDC)Dptztz] crystallizes in the triclinic space group P-1. The asymmetric unit consists of one Zn(II) ion, one molecule of the linkers 1,3-BDC^{2-} and Dptztz, each, and a dimethylformamide (DMF) solvent molecule (Figure 4, Figure S2). One carboxylate group of 1,3-BDC^{2-} connects two symmetry equivalent Zn(II) ions in a syn-syn-bis-monodentate coordination mode into a dinuclear unit with a Zn$\cdots$Zn distance of 4.082 Å. The other carboxylate group chelates an adjacent Zn atom. Thereby the 1,3-BDC linkers bridge between neighboring dinuclear entities to form a one-dimensional double strand along the *b*-axis (Figure 5a). These double strands are pillared by Dptztz into a 2D coordination network structure (Figure 5b). The secondary building unit of the structure is the dinuclear unit {Zn$_2$(O$_2$C)$_4$N$_4$}. The 2D layers assemble through π-π interactions between isophthalate aryl rings and CH-π interactions between an isophthalate and a pyridyl-moiety of Dptztz of adjacent layers into a 3D supramolecular network (Figure 5c and Figure S3, Table S1). The 2D network in [Zn(1,3-BDC)Dptztz] is isotopic to the aforementioned CIDs (coordination polymers with interdigitated structure) studied by Kitagawa and co-workers [24–29].

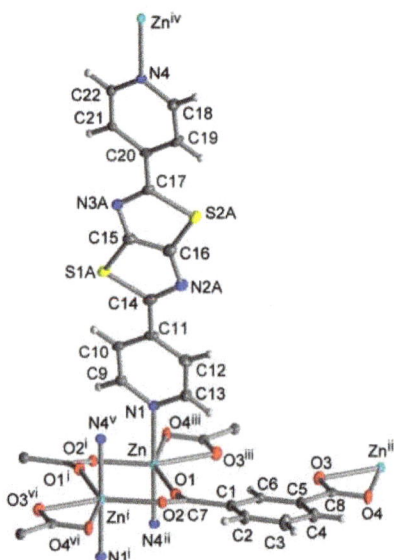

Figure 4. Extended asymmetric unit of [Zn(1,3-BDC)Dptztz]·DMF (50% thermal ellipsoids; symmetry transformations: i = −x + 2, −y, −z + 1; ii = x + 1, y, z + 1; iii = −x + 2, −y + 1, −z + 1; iv = x − 1, y, z − 1; v = −x + 1, −y, −z, vi = x, y − 1, z). For the slight ring-flip disorder of the thiazolothiazol moiety and the DMF solvent molecule, which is omitted here for clarity, see Figure S2 in Supplementary Material. See Table 2 for selected bond lengths and angles.

Table 2. Selected bond lengths and angles (Å, °) in [Zn(1,3-BDC)Dptztz].

Zn–O1	2.0532 (18)	Zn–O4^{iii}	2.2269 (19)
Zn–O2^{i}	2.0218 (18)	Zn–N1	2.166 (3)
Zn–O3^{iii}	2.1569 (19)	Zn–N4^{ii}	2.151 (3)
O1–Zn–O2^{i}	119.30 (7)	O2^{i}–Zn–N4^{ii}	89.34 (9)
O1–Zn–O3^{iii}	88.78 (7)	O3^{iii}–Zn–O4^{iii}	60.01 (7)
O1–Zn–O4^{iii}	148.39 (7)	O3^{iii}–Zn–N1	90.18 (9)
O1–Zn–N1	89.84 (9)	O3^{iii}–Zn–N4^{ii}	90.56 (9)
O1–Zn-N4^{ii}	86.00 (9)	O4^{iii}–Zn–N1	94.77 (8)
O2^{i}–Zn–O3^{iii}	151.83 (7)	O4^{iii}–Zn–N4^{ii}	89.21 (8)
O2^{i}–Zn–O4^{iii}	91.82 (7)	N1–Zn–N4^{ii}	175.76 (8)
O2^{i}–Zn–N1	91.99 (9)		

Symmetry transformations: i = −x + 2, −y, −z + 1; ii = x + 1, y, z + 1; iii = −x + 2, −y + 1, −z + 1.

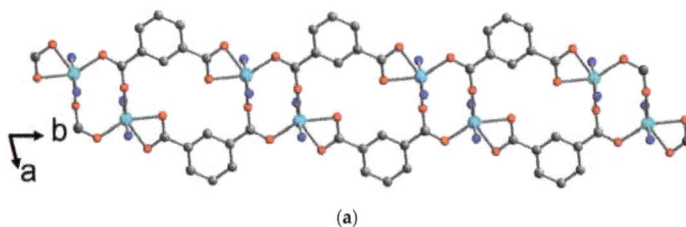

(a)

Figure 5. *Cont.*

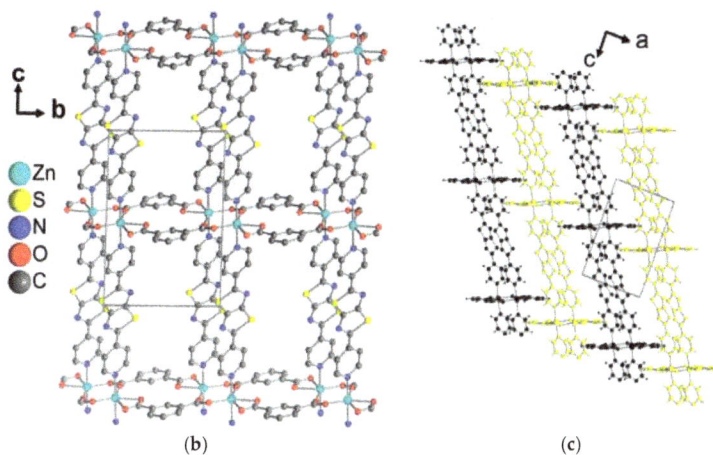

(b)   (c)

Figure 5. (a) 1D double strand of Zn^{2+} and 1,3-BDC^{2-} along the *b*-axis and (b) 2D coordination network structure in the *bc* plane and (c) supramolecular 3D packing of the 2D layers in [Zn(1,3-BDC)Dptztz]·DMF (H atoms in (**a,b**) and DMF solvent molecules are not shown for clarity). In (c) the 2D layers are colored alternately black and yellow for clarity.

The thermogravimetric analysis in Figure 6 shows a weight loss of 10.6% between 90 and 200 °C (calc. 12.2% for one DMF molecule per formula unit of [Zn(1,3-BDC)Dptztz]·DMF). The residual [Zn(1,3-BDC)Dptztz] framework shows stability up to 280 °C.

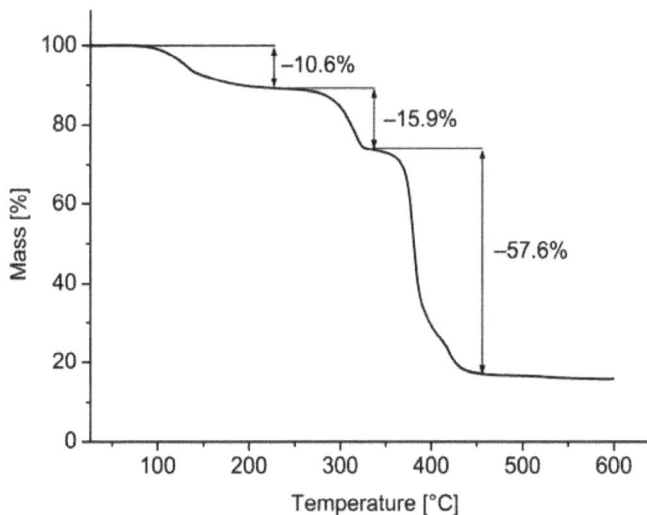

Figure 6. TGA curve of [Zn(1,3-BDC)Dptztz]·DMF·in the temperature range 26–600 °C with a heating rate of 10 K/min under N$_2$ atmosphere.

Prior to the sorption experiments the crystals of [Zn(1,3-BDC)Dptztz]·DMF were collected by suction filtration. Afterwards they were suspended in acetonitrile for three days to induce solvent exchange as part of the sample activation procedure. Additionally, the acetonitrile solvent was exchanged three times per day. Afterwards, the sample was degassed at 120 °C for 15 h under vacuum.

The activated compound [Zn(1,3-BDC)Dptztz] shows no uptake of N_2 at 77 K, which is in accordance with the observations by Kitagawa et al. for analogous CID structures [24–29]. For CO_2—with its large polarizability and quadrupole moment—[Zn(1,3-BDC)Dptztz] shows a type I adsorption isotherm at 195 K with a maximum uptake of 138 cm^3/g at 753 mmHg of CO_2 (Figure 7). At higher absolutes pressures the desorption curve shows a small hysteresis, but at low pressures the hysteresis gap closes. This proves the microporous nature of [Zn(1,3-BDC)Dptztz]. Because [Zn(1,3-BDC)Dptztz] is non-porous towards N_2 the CO_2 data was used to calculated the BET surface area. The BET surface area from the CO_2 adsorption isotherm is 417 m^2/g (calculated from p/p_0 = 0.00–0.04). The pore volume is 0.246 cm^3/g at p/p_0 = 0.5 calculated from the isotherm measured at 195 K. The calculated accessible surface area is 25.6% or 0.185 cm^3/g calculated with PLATON from the DMF solvent-depleted structure. The measured pore volume is about 35% higher than the one calculated from the crystal structure data. This can be interpreted such that CO_2 can create a larger interlayer volume through strong interaction with the highly polarized surface area at 195 K.

Additionally, the CO_2 isotherms at 273 K and 293 K were measured (Figure 7). The data is presented in Table 3. From the CO_2 isotherms at 273 K and 293 K, the heat of adsorption at zero coverage was derived as 26.2 kJ/mol. The heat of adsorption of CO_2 for MOFs can vary from 20 kJ/mol to over 90 kJ/mol. A higher heat of adsorption is usually indicative for stronger framework CO_2 interactions [47]. Representative values are 20kJ/mol for [Zn(1,4-BDC)(TED)], 30 kJ/mol for HKUST-1, 47 kJ/mol for Mg-MOF-74, 63 kJ/mol for MIL-100-Cr and 96 kJ/mol for mmen-Cu-BTTri [48–52].

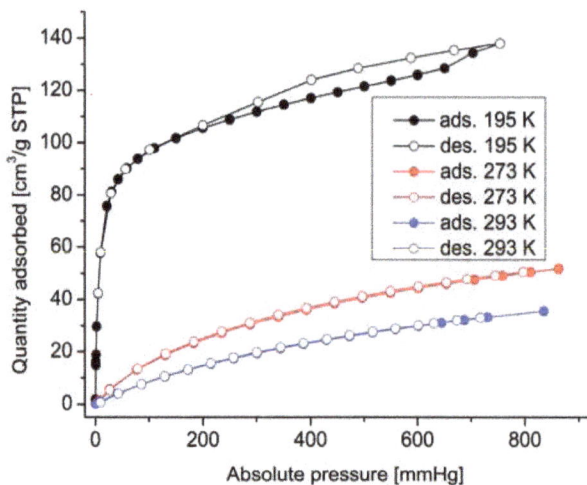

Figure 7. CO_2 adsorption (closed symbols) and desorption (open symbols) isotherms for activated [Zn(1,3-BDC)Dptztz] measured at 195 K (**black**), 293 K (**red**) and 273 K (**blue**).

Table 3. CO_2 sorption data for [Zn(1,3-BDC)Dptztz].

	Quantity Adsorbed (cm^3/g, mmol/g, wt %)	Total Pore Volume (cm^3/g)
195 K	138, 6.16, 27.1%	0.246 [1]
273 K	51.9, 2.32, 10.2%	0.092 [2]
293 K	35.5, 1.59, 7.0%	0.061 [3]

[1] at p/p_0 = 0.50, [2] at p/p_0 = 0.03, [3] at p/p_0 = 0.017.

The heat of adsorption curve (Figure 8) has a relative maximum at a quantity adsorbed of about 4 cm^3/g STP with 27.7 kJ/mol. Afterwards the heat of adsorption decreases to 25.7 kJ/mol. For most MOF materials, the heat of adsorption curve decreases monotonically, since the adsorption sites with the highest affinity towards the adsorbate are occupied first and at higher loadings the adsorption sites usually have weaker affinities. Two MOF classes for which the heat of adsorption does not decrease monotonically are the MIL-53 and MIL-47 series. Férey et al. suggested that the transformation of MIL-53 from a closed or narrow-pore to the open or wide-pore phase is an endothermic process and that this process is balanced with the exothermic adsorption process. Subsequently, the MIL-53 MOFs also have a local maximum between 0 and 1 bar (see Figure S4 in Supplementary Material) [53]. Many CIDs also show gate-opening phenomena and/or an adsorbate specific expansion upon the adsorption process. Yet, to the best of our knowledge, no heat of adsorption curves for CIDs are published in the literature. But Pera–Titus and Farruseng calculated the phase transition energies for CID-21 and CID-22 (benzene and tetrazine spacer) to be 1.4 and 1.5 kJ/mol from the closed to the open phase for CO_2 adsorption at 195 K, respectively [54]. These values are in good accordance with the difference between the heat of adsorption at zero coverage and the relative maximum of the curve in Figure 8. So it can be concluded that [Zn(1,3-BDC)Dptztz] shows a gate-opening effect for CO_2, which is represented in a non-monotonic heat of adsorption curve.

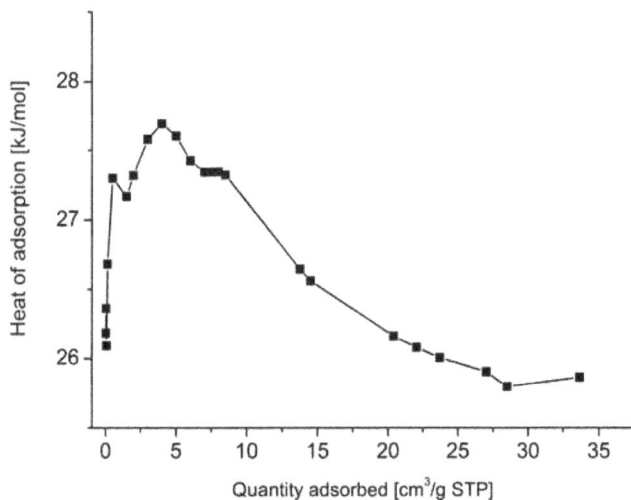

Figure 8. Heat of adsorption plot of CO_2 adsorption for [Zn(1,3-BDC)Dptztz].

Compound [Zn(1,3-BDC)Dptztz] gradually adsorbs H_2O at 293 K with a maximum uptake of 121 mg/g at 0.9 p/p$_0$ (Figure 9). This uptake equals 3.5 H_2O molecules per asymmetric unit. The desorption curve shows a hysteresis, indicative for a strong interaction of H_2O with the framework. [Zn(1,3-BDC)Dptztz] adsorbs CO_2 at 195 K due to its large polarizability and quadrupole moment while N_2 at 77 K is not adsorbed. The adsorption characteristics of [Zn(1,3-BDC)Dptztz] apparently depend on the interaction between the adsorbate and the framework und not only on the pore size. The sorption characteristics of [Zn(1,3-BDC)Dptztz] towards H_2O with its pronounced hysteresis underpin these findings. It can be concluded that the decoration of the surface with the polarized and electron poor tztz moiety can alter sorption properties to become highly selective towards polarizable adsorbents. Further studies on different tztz-containing materials are underway in our institute.

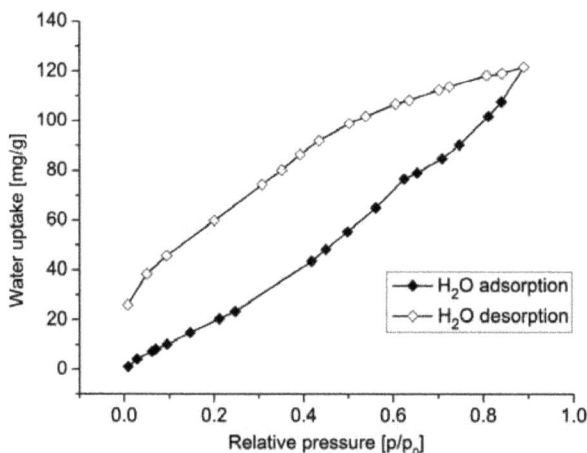

Figure 9. Water sorption isotherms at 293 K for [Zn(1,3-BDC)Dptztz].

4. Conclusions

The 4,4′-dipyridyl *N,N′*-donor ligand Dptztz with the central thiazolo[5,4-d]thiazole unit was successfully synthesized and its crystal structure was determined for the first time. With the linker Dptztz, the thiazolo[5,4-d]thiazole-unit was integrated into a solvent-filled coordination network of the formula [Zn(1,3-BDC)Dptztz]·DMF belonging to the class of coordination polymers with interdigitated structure (CIDs). Synthesis of the coordination network was carried out via a mixed ligand strategy in a solvothermal reaction. Interdigitation between the 2D layers to a 3D supramolecular network appears to be controlled by π-π interactions between isophthalate aryl rings and CH-π interactions between isophthalate and pyridyl moieties. Activated [Zn(1,3-BDC)Dptztz] shows a BET surface of 417 m^2/g calculated from CO_2 adsorption data, while N_2 which unlike CO_2 is not as polarizable and has no quadrupole moment is not adsorbed. The heat of adsorption for CO_2 exhibits a relative maximum at a quantity adsorbed of about 4 cm^3/g STP with 27.7 kJ/mol, which is interpreted as a gate-opening effect. This is the first report of the sorption characteristics of a tztz-functionalized porous MOF material.

Supplementary Materials: The following are available online at www.mdpi.com/2073-4352/8/1/30/s1. Figure S1: Section of the packing diagram of 2,5-di(4-pyridyl)thiazolo[5,4-d]thiazole (Dptztz), showing (a) the complementary C1-H1···N1 hydrogen bonds and (b) the π-π interactions, thereby giving a supramolecular layer in the *ab* plane; Figure S2: Asymmetric unit of [Zn(1,3-BDC)Dptztz]·DMF showing the ring-flip disorder of the thiazolothiazol moiety; Figure S3: π-π and CH-π interactions between the 2D layers in [Zn(1,3-BDC)Dptztz]; Figure S4: Heat of adsorption curve of MIL-53-Cr; Table S1. Distances [Å] and angles [°] for the shortest π-π and CH-π supramolecular interactions between the 2D layers in [Zn(1,3-BDC)Dptztz].

Acknowledgments: The work was supported by the Federal German Ministry of Education and Research (BMBF) under grant-# 03SF0492C (Optimat).

Author Contributions: Simon Millan designed the experiments, synthesized the ligand and [Zn(1,3-BDC)Dptztz]. Gamall Makhloufi collected the single crystal data. Data analysis and measurements were performed by Simon Millan, while Christoph Janiak and Simon Millan wrote the manuscript.

Conflicts of Interest: The authors declare no conflict of interest. The founding sponsors had no role in the design of the study; in the collection, analyses, or interpretation of data; in the writing of the manuscript, and in the decision to publish the results.

References

1. Yaghi, O.M.; O'Keeffe, M.; Ockwig, N.W.; Chae, H.K.; Eddaoudi, M.; Kim, J. Reticular synthesis and the design of new materials. *Nature* **2003**, *423*, 705–714. [CrossRef] [PubMed]

2. Maurin, G.; Serre, C.; Cooper, A.; Férey, G. The new age of MOFs and of their porous-related solids. *Chem. Soc. Rev.* **2017**, *46*, 3104–3107. [CrossRef] [PubMed]

3. Furukawa, H.; Ko, N.; Go, Y.B.; Aratani, N.; Choi, S.B.; Choi, E.; Yazaydin, A.Ö.; Snurr, R.Q.; O'Keeffe, M.; Kim, J.; Yaghi, O.M. Ultrahigh porosity in metal-organic frameworks. *Science* **2010**, *329*, 424–428. [CrossRef] [PubMed]

4. Adil, K.; Belmabkhout, Y.; Pillai, R.S.; Cadiau, A.; Bhatt, P.M.; Assen, A.H.; Maurin, G.; Eddaoudi, M. Gas/vapour separation using ultra-microporous metal–organic frameworks: Insights into the structure/separation relationship. *Chem. Soc. Rev.* **2017**, *46*, 3402–3430. [CrossRef] [PubMed]

5. Li, J.-R.; Kuppler, R.J.; Zhou, H.-C. Selective gas adsorption and separation in metal-organic frameworks. *Chem. Soc. Rev.* **2009**, *38*, 1477–1504. [CrossRef] [PubMed]

6. Dechnik, J.; Gascon, J.; Doonan, C.J.; Janiak, C.; Sumby, C.J. Mixed-matrix membranes. *Angew. Chem. Int. Ed.* **2017**, *56*, 9292–9310. [CrossRef] [PubMed]

7. Lee, J.; Farha, O.K.; Roberts, J.; Scheidt, K.A.; Nguyen, S.T.; Hupp, J.T. Metal-organic framework materials as catalysts. *Chem. Soc. Rev.* **2009**, *38*, 1450–1459. [CrossRef] [PubMed]

8. Herbst, A.; Janiak, C. MOF catalysts in biomass upgrading towards value-added fine chemicals. *CrystEngComm* **2017**, *19*, 4092–4117. [CrossRef]

9. Kitao, T.; Zhang, Y.; Kitagawa, S.; Wang, B.; Uemura, T. Hybridization of MOFs and polymers. *Chem. Soc. Rev.* **2017**, *46*, 3108–3133. [CrossRef] [PubMed]

10. Lustig, W.P.; Mukherjee, S.; Rudd, N.D.; Desai, A.V.; Li, J.; Ghosh, S.K. Metal–organic frameworks: functional luminescent and photonic materials for sensing applications. *Chem. Soc. Rev.* **2017**, *46*, 3242–3285. [CrossRef] [PubMed]

11. Hu, Z.; Deibert, B.J.; Li, J. Luminescent metal-organic frameworks for chemical sensing and explosive detection. *Chem. Soc. Rev.* **2014**, *43*, 5815–5840. [CrossRef] [PubMed]

12. Gangu, K.K.; Maddila, S.; Mukkamala, S.B.; Jonnalagadda, S.B. A review on contemporary Metal–Organic Framework materials. *Inorg. Chim. Acta* **2016**, *446*, 61–74. [CrossRef]

13. Jeremias, F.; Fröhlich, D.; Janiak, C.; Henninger, S.K. Water and methanol adsorption on MOFs for cycling heat transformation processes. *New J. Chem.* **2014**, *38*, 1846–1852. [CrossRef]

14. Almeida Paz, F.A.; Klinowski, J.; Vilela, S.M.F.; Tome, J.P.C.; Cavaleiro, J.A.S.; Rocha, J. Ligand design for functional metal-organic frameworks. *Chem. Soc. Rev.* **2012**, *41*, 1088–1110. [CrossRef] [PubMed]

15. Tanabe, K.K.; Cohen, S.M. Postsynthetic modification of metal-organic frameworks—A progress report. *Chem. Soc. Rev.* **2011**, *40*, 498–519. [CrossRef] [PubMed]

16. Islamoglu, T.; Goswami, S.; Li, Z.; Howarth, A.J.; Farha, O.K.; Hupp, J.T. Postsynthetic tuning of metal–organic frameworks for targeted applications. *Acc. Chem. Res.* **2017**, *50*, 805–813. [CrossRef] [PubMed]

17. Biradha, K.; Sarkar, M.; Rajput, L. Crystal engineering of coordination polymers using 4,4′-bipyridine as a bond between transition metal atoms. *Chem. Commun.* **2006**, 4169–4179. [CrossRef] [PubMed]

18. Bhattacharya, B.; Ghoshal, D. Selective carbon dioxide adsorption by mixed-ligand porous coordination polymers. *CrystEngComm* **2015**, *17*, 8388–8413. [CrossRef]

19. Haldar, R.; Maji, T.K. Metal-organic frameworks (MOFs) based on mixed linker systems: Structural diversities towards functional materials. *CrystEngComm* **2013**, *15*, 9276–9295. [CrossRef]

20. Bae, Y.-S.; Mulfort, K.L.; Frost, H.; Ryan, P.; Punnathanam, S.; Broadbelt, L.J.; Hupp, J.T.; Snurr, R.Q. Separation of CO_2 from CH_4 using mixed-ligand metal–organic frameworks. *Langmuir* **2008**, *24*, 8592–8598. [CrossRef] [PubMed]

21. Henke, S.; Schneemann, A.; Wütscher, A.; Fischer, R.A. Directing the breathing behavior of pillared-layered metal–organic frameworks via a systematic library of functionalized linkers bearing flexible substituents. *J. Am. Chem. Soc.* **2012**, *134*, 9464–9474. [CrossRef] [PubMed]

22. Glomb, S.; Woschko, D.; Makhloufi, G.; Janiak, C. Metal–organic frameworks with internal urea-functionalized dicarboxylate linkers for SO_2 and NH_3 adsorption. *ACS Appl. Mater. Interfaces* **2017**, *9*, 37419–37434. [CrossRef] [PubMed]

23. Takashima, Y.; Martínez, V.M.; Furukawa, S.; Kondo, M.; Shimomura, S.; Uehara, H.; Nakahama, M.; Sugimoto, K.; Kitagawa, S. Molecular decoding using luminescence from an entangled porous framework. *Nat. Commun.* **2011**, *2*, 168. [CrossRef] [PubMed]

24. Horike, S.; Tanaka, D.; Nakagawa, K.; Kitagawa, S. Selective guest sorption in an interdigitated porous framework with hydrophobic pore surfaces. *Chem. Commun.* **2007**, 3395–3397. [CrossRef]
25. Tanaka, D.; Nakagawa, K.; Higuchi, M.; Horike, S.; Kubota, Y.; Kobayashi, T.C.; Takata, M.; Kitagawa, S. Kinetic gate-opening process in a flexible porous coordination polymer. *Angew. Chem. Int. Ed.* **2008**, *47*, 3914–3918. [CrossRef] [PubMed]
26. Fukushima, T.; Horike, S.; Inubushi, Y.; Nakagawa, K.; Kubota, Y.; Takata, M.; Kitagawa, S. Solid solutions of soft porous coordination polymers: Fine-tuning of gas adsorption properties. *Angew. Chem.* **2010**, *122*, 4930–4934. [CrossRef]
27. Nakagawa, K.; Tanaka, D.; Horike, S.; Shimomura, S.; Higuchi, M.; Kitagawa, S. Enhanced selectivity of CO_2 from a ternary gas mixture in an interdigitated porous framework. *Chem. Commun.* **2010**, *46*, 4258–4260. [CrossRef] [PubMed]
28. Hijikata, Y.; Horike, S.; Sugimoto, M.; Sato, H.; Matsuda, R.; Kitagawa, S. Relationship between channel and sorption properties in coordination polymers with interdigitated structures. *Chem. Eur. J.* **2011**, *17*, 5138–5144. [CrossRef] [PubMed]
29. Kishida, K.; Horike, S.; Nakagawa, K.; Kitagawa, S. Synthesis and adsorption properties of azulene-containing porous interdigitated framework. *Chem. Lett.* **2012**, *41*, 425–426. [CrossRef]
30. Johnson, J.R.; Ketcham, R. Thiazolothiazoles. I. The reaction of aromatic aldehydes with dithiooxamide. *J. Am. Chem. Soc.* **1960**, *82*, 2719–2724. [CrossRef]
31. Bevk, D.; Marin, L.; Lutsen, L.; Vanderzande, D.; Maes, W. Thiazolo[5,4-d]thiazoles - promising building blocks in the synthesis of semiconductors for plastic electronics. *RSC Adv.* **2013**, *3*, 11418–11431. [CrossRef]
32. Reginato, G.; Mordini, A.; Zani, L.; Calamante, M.; Dessi, A. Photoactive compounds based on the thiazolo[5,4-d]thiazole core and their application in organic and hybrid photovoltaics. *Eur. J. Org. Chem.* **2016**, 233–251. [CrossRef]
33. Zampese, J.A.; Keene, F.R.; Steel, P.J. Diastereoisomeric dinuclear ruthenium complexes of 2,5-di(2-pyridyl)thiazolo[5,4-d]thiazole. *Dalton Trans.* **2004**, 4124–4129. [CrossRef] [PubMed]
34. Falcão, E.H.L.; Naraso; Feller, R.K.; Wu, G.; Wudl, F.; Cheetham, A.K. Hybrid organic–inorganic framework structures: Influence of cation size on metal–oxygen–metal connectivity in the alkaline earth thiazolothiazoledicarboxylates. *Inorg. Chem.* **2008**, *47*, 8336–8342.
35. Aprea, A.; Colombo, V.; Galli, S.; Masciocchi, N.; Maspero, A.; Palmisano, G. Thiazolo[5,4-d]thiazole-2,5-dicarboxylic acid, $C_6H_2N_2O_4S_2$, and its coordination polymers. *Solid State Sci.* **2010**, *12*, 795–802. [CrossRef]
36. Rizzuto, F.J.; Faust, T.B.; Chan, B.; Hua, C.; D'Alessandro, D.M.; Kepert, C.J. Experimental and computational studies of a multi-electron donor–acceptor ligand containing the thiazolo[5,4-d]thiazole core and its incorporation into a metal–organic framework. *Chem. Eur. J.* **2014**, *20*, 17597–17605. [CrossRef] [PubMed]
37. Hua, C.; Rizzuto, F.J.; Zhang, X.; Tuna, F.; Collison, D.; D'Alessandro, D.M. Spectroelectrochemical properties of a Ru(II) complex with a thiazolo[5,4-d]thiazole triarylamine ligand. *New J. Chem.* **2017**, *41*, 108–114. [CrossRef]
38. Zhu, X.; Tian, C.; Jin, T.; Wang, J.; Mahurin, S.M.; Mei, W.; Xiong, Y.; Hu, J.; Feng, X.; Liu, H. Thiazolothiazole-linked porous organic polymers. *Chem. Commun.* **2014**, *50*, 15055–15058. [CrossRef] [PubMed]
39. APEX2. *SAINT, Data Reduction and Frame Integration Program for the CCD Area-Detector System, Bruker Analytical X-ray Systems*; Data Collection Program for the CCD Area-Detector System: Madison, WI, USA, 1997–2006.
40. Sheldrick, G.M. *SADABS: Area-Detector Absorption Correction*; University of Göttingen: Göttingen, Germany, 1996.
41. Sheldrick, G.M. Crystal structure refinement with SHElXL. *Acta Crystallogr. Sect A* **2015**, *71*, 3–8. [CrossRef] [PubMed]
42. Sheldrick, G. A short history of SHELX. *Acta Crystallogr. Sect. A* **2008**, *64*, 112–122. [CrossRef] [PubMed]
43. Brandenburg, K. *DIAMOND*, version 4.4; Crystal and Molecular Structure Visualization; Crystal Impact—K. Brandenburg & H. Putz Gbr: Bonn, Germany, 2009–2017.
44. Spek, A.L. Structure validation in chemical crystallography. *Acta Crystallogr. Sect. D—Biol. Crystallogr.* **2009**, *65*, 148–155. [CrossRef] [PubMed]
45. Knighton, R.C.; Hallett, A.J.; Kariuki, B.M.; Pope, S.J.A. A one-step synthesis towards new ligands based on aryl-functionalised thiazolo[5,4-d]thiazole chromophores. *Tetrahedron Lett.* **2010**, *51*, 5419–5422. [CrossRef]

46. Hisamatsu, S.; Masu, H.; Azumaya, I.; Takahashi, M.; Kishikawa, K.; Kohmoto, S. U-shaped aromatic ureadicarboxylic acids as versatile building blocks: Construction of ladder and zigzag networks and channels. *Cryst. Growth Des.* **2011**, *11*, 5387–5395. [CrossRef]

47. Das, A.; D'Alessandro, D.M. Tuning the functional sites in metal-organic frameworks to modulate CO_2 heats of adsorption. *CrystEngComm* **2015**, *17*, 706–718. [CrossRef]

48. Zhao, Y.; Wu, H.; Emge, T.J.; Gong, Q.; Nijem, N.; Chabal, Y.J.; Kong, L.; Langreth, D.C.; Liu, H.; Zeng, H.; Li, J. Enhancing gas adsorption and separation capacity through ligand functionalization of microporous metal–organic framework structures. *Chem. Eur. J.* **2011**, *17*, 5101–5109. [CrossRef] [PubMed]

49. Liang, Z.; Marshall, M.; Chaffee, A.L. CO_2 adsorption-based separation by metal organic framework (Cu-BTC) versus zeolite (13X). *Energy Fuels* **2009**, *23*, 2785–2789. [CrossRef]

50. Caskey, S.R.; Wong-Foy, A.G.; Matzger, A.J. Dramatic tuning of carbon dioxide uptake via metal substitution in a coordination polymer with cylindrical pores. *J. Am. Chem. Soc.* **2008**, *130*, 10870–10871. [CrossRef] [PubMed]

51. Llewellyn, P.L.; Bourrelly, S.; Serre, C.; Vimont, A.; Daturi, M.; Hamon, L.; De Weireld, G.; Chang, J.-S.; Hong, D.-Y.; Kyu Hwang, Y.; et al. High uptakes of CO_2 and CH_4 in mesoporous metal-organic frameworks MIL-100 and MIL-101. *Langmuir* **2008**, *24*, 7245–7250. [CrossRef] [PubMed]

52. McDonald, T.M.; D'Alessandro, D.M.; Krishna, R.; Long, J.R. Enhanced carbon dioxide capture upon incorporation of *N,N'*-dimethylethylenediamine in the metal-organic framework CuBTTri. *Chem. Sci.* **2011**, *2*, 2022–2028. [CrossRef]

53. Bourrelly, S.; Llewellyn, P.L.; Serre, C.; Millange, F.; Loiseau, T.; Férey, G. Different adsorption behaviors of methane and carbon dioxide in the isotypic nanoporous metal terephthalates MIL-53 and MIL-47. *J. Am. Chem. Soc.* **2005**, *127*, 13519–13521. [CrossRef] [PubMed]

54. Pera-Titus, M.; Farrusseng, D. Guest-induced gate opening and breathing phenomena in soft porous crystals: Building thermodynamically consistent isotherms. *J. Phys. Chem. C* **2012**, *116*, 1638–1649. [CrossRef]

crystals

MDPI

Communication

Synthesis and Crystal Structure of a Zn(II)-Based MOF Bearing Neutral N-Donor Linker and SiF_6^{2-} Anion

Biplab Manna [1,†], Shivani Sharma [1,†] and Sujit K. Ghosh [1,2,*]

[1]   Department of Chemistry, Indian Institute of Science Education and Research (IISER Pune),
      Dr. Homi Bhabha Road, Pashan, Pune 411008, India; biplabchem.manna@gmail.com (B.M.);
      shivani.sharma@students.iiserpune.ac.in (S.S.)
[2]   Centre for Research in Energy & Sustainable Materials, IISER Pune, Dr. Homi Bhabha Road, Pashan,
      Pune 411008, India
*   Correspondence: sghosh@iiserpune.ac.in; Tel.: +91-20-2590-8076
†   These authors contributed equally to this work.

Received: 20 November 2017; Accepted: 10 January 2018; Published: 16 January 2018

Abstract: A novel three-dimensional two-fold interpenetrated bi-porous metal-organic framework IPM-325 (IPM: IISER Pune Materials) having **pcu** topology was synthesized at room temperature. Single crystal X-ray diffraction (SC-XRD) study revealed that the compound crystallizes in monoclinic lattice with molecular formula $\{[Zn(L)_2 (SiF_6)] (CH_2Cl_2) xG\}_n$ where G = Guests). All metal centers were found to have octahedral geometry. From single crystal analysis it can be inferred that SiF_6^{2-} anion play a vital role in extending the dimensionality of the framework by bridging between two metal centers. Interestingly, IPM-325 exhibited two-step structural transformation maintaining the crystallinity of the framework as characterized by powder X-ray diffraction (PXRD).

Keywords: metal-organic framework; neutral N-donor framework; structural transformations

---

1. Introduction

Metal-organic frameworks (MOFs)/Porous coordination polymers (PCPs) have shown immense potential in various domains including separation, storage, sensing, catalysis, etc. [1–5]. MOFs are extended networks constituting of metal nodes and organic moiety linked via coordination bonds resulting in framework formation with potential voids [6,7]. MOFs can be readily functionalized using different organic building blocks thus imparting flexibility to tune the properties [8]. Metal-organic frameworks can also be tuned by using varied metals/metal clusters along with counter ions with different coordinating tendency [9–13]. MOFs have been classified as neutral and ionic MOFs (iMOFs) based on the framework charge [14]. Among various type of linkers, neutral nitrogen donor-based linkers have played a vital role since the emergence of MOFs owing to the easy availability and facile coordinating ability [15,16]. In pioneering reports, Zaworotko and Kitagawa have reported MOFs utilizing SiF_6^{2-} as counter anion and neutral nitrogen donor linker [17,18]. Recently, (SiF_6^{2-}) anion has been utilized extensively due to the tendency of SiF_6^{2-} to bridge two-dimensional sheets to form an overall three-dimensional framework with formation of one dimensional channel with varied porosity depending on the length of the linker [19,20]. Owing to the presence of highly electronegative fluorine atoms in SiF_6^{2-} it leads to the formation of highly charged polar surfaces. Owing to the highly charged polar surfaces, SiF_6^{2-} anion based MOFs are currently being pursued in various sorption-based applications like CO_2 capture, water sorption, hydrocarbon separation, etc. [21–23]. Thus, SiF_6^{2-} as an inorganic pillar should be utilized to develop novel materials as it imparts higher selectivity in sorption-based applications.

Yet another important classification within the field of MOFs is to classify MOFs as rigid MOF or flexible MOF depending on the frameworks tendency to change the structure on the application of varied stimuli like pressure, temperature, etc. [24–26]. Flexible MOFs are the frameworks which tend to change the overall structure upon applications of different stimuli. Flexible MOFs have been utilized in various adsorption-based separation processes with promising results [27,28]. Thus, development of MOFs that exhibit structural transformation is quite crucial for the development of materials exhibiting selective sorption.

SiF_6^{2-} anion-based rigid MOFs have been extensively studied but less attention has been paid towards flexible SiF_6^{2-}-based MOF system. SiF_6^{2-} anion-based flexible MOFs may show dual facet of high selectivity, hysteretic sorption along with various characteristics of flexible MOFs accompanied with properties imparted by SiF_6^{2-} anions. Herein, we report the crystal structure of IPM-325 which exhibits structural transformations as characterized by PXRD.

2. Materials and Methods

All the starting reagents zinc hexafluorosilicate hydrate, (1*E*,2*E*)-1,2-bis(pyridin-4-ylmethylene) hydrazine were purchased from Sigma-Aldrich, Bangalore Karnataka, India and all the solvents were procured locally. Powder X-ray diffraction (PXRD) patterns were measured on Bruker D8 Advanced X-ray diffractometer (Bruker AXS GmbH, Karlsruhe, Germany) at room temperature using Cu-Kα radiation (λ = 1.5406 Å) with a scan speed of 0.5° min^{-1} and a step size of 0.01° in 2 *theta*. Thermogravimetric analysis results were obtained in the temperature range of 30–800 °C on Perkin-Elmer STA 6000 analyzer (TGA, STA 6000 Perkin Elmer, PerkinElmer, Inc., Waltham, MA, USA) under N$_2$ atmosphere, at a heating rate of 10 °C min^{-1}. Single-crystal X-ray data of IPM-325_P1 was collected at 100 K on a Bruker KAPPA APEX II CCD Duo diffractometer (operated at 1500 W power: 50 kV, 30 mA, Bruker AXS GmbH, Karlsruhe, Germany) using graphite-monochromated Mo Kα adiation (λ = 0.71073 Å). A crystal was mounted on a nylon Cryoloop (Hampton Research) with Paraton-N oil (Hampton Research). The data integration and reduction were processed with SAINT software [29]. A multi-scan absorption correction was applied to the collected reflections [30]. The structure was solved by the direct method using SHELXS-2014 [31] and was refined on *F2* by full-matrix least-squares technique using the SHELXL-2017 [32] program package within the WINGX program [33]. All non-hydrogen atoms were refined anisotropically. All hydrogen atoms were located in successive difference Fourier maps and they were treated as riding atoms using SHELXL default parameters. Highly disordered guest solvent molecules were observed. SQUEEZE option was used that eliminate the contribution of disordered guest molecules [34]. Residual electron density was calculated from the SQUEEZE function 49 electrons/unit cell. Gas adsorption measurements were performed using BelSorp-Max instrument (BEL Japan, Inc., Osaka, Japan).

Experimental Synthesis

Synthesis of Linker L ((1*E*,2*E*)-1,2-bis(pyridin-4-ylmethylene)hydrazine): Linker L was synthesized by modifying a reported protocol [35]. 4-pyridinecarboxaldehyde (49.94 mmol, 4.704 mL) was taken in a round bottomed flask. To this flask (1:1) Methanol (15 mL) and ethanol (15 mL) solution was added and catalytic quantity of acetic acid was added to this mixture. This mixture was stirred at 120 °C for 0.5 h. Further, hydrazine hydrate (19.97 mmol, 0.970 mL) was added and the reaction was maintained at 120 °C for 12 h. Subsequent cooling to room temperature yielded yellow crystalline product. Crude product was recrystallized from a solvent combination of methanol: ethanol solvent mixture. Yield = 90%.

Synthesis of IPM-325_p1: Single crystals of IPM-325_p1 were obtained by slow diffusion 1:1 mixture of zinc hexafluorosilicate hydrate (ZnSiF$_6$, H$_2$O) (0.1 mmol, 20.7 mg) and L = (1*E*,2*E*)-1,2-bis(pyridin-4-ylmethylene)hydrazine (0.1 mmol, 21 mg) in a solvent combination of Dichloromethane, methanol and benzene. Block shaped yellow colored crystal was obtained after one week. Yield ~60%.

Desolvated phase preparation: IPM-325_p1 was heated at 160 °C overnight with slow increase in the temperature at the rate of 5 °C/h and then maintaining 160 °C overnight. The desolvated phase was further characterized by thermogravimetric analysis.

3. Results and Discussions

IPM-325 was synthesized by slow diffusion of ligand L (Scheme 1) and $ZnSiF_6$ at room temperature to yield block shaped yellow colored crystals. Single crystal X-ray diffraction analysis revealed that the compound crystallizes in *C2/m* space group with Z = 2 (Table 1). The asymmetric unit consist of Zn(II) metal ion along with two linkers and SiF_6^{2-} anion with dichloromethane molecules and other disordered molecules present within the voids (Figure 1). Zn(II) metal node is octahedrally coordinated with N_4F_2 donor set with the distance between Zn(II) and equatorial nitrogen atoms 2.139 Å while the distance between Zn(II) and axial F atoms was 2.153 Å. SiF_6^{2-} anions were found to bridge between two metal centers (Figure 2a). Overall structure was found to be two-fold interpenetrated in nature with an overall **pcu** topology (Figure 2b). Along the *a*-axis, the packing diagram showed types of rectangular voids Pore A and Pore B having dimensions 5.55×7.46 Å^2 and 5.84×6.54 Å^2 respectively (excluding van der Waals radii) (Figure 2b). The voids of the asynthesized framework (Figure 3) (IPM-325_p1; p represents phase) are filled with dichloromethane solvent wherein the H6 atom exhibits hydrogen bonding interaction with the F3 atom present at the metal node. To establish the bulk phase purity of IPM-325_p1, it was characterized using powder X-Ray diffraction measurements (PXRD) which showed a considerable shift in the observed pattern when compared to the simulated pattern (Figure 4a. The observed shift may be due to the overall structural transformation leading to formation of new phase named as IPM-325_p2 (p represents phase). To establish the bulk phase purity, we performed PXRD analysis of asynthesized crystals in presence of mother liquor. The PXRD pattern corroborated well with the simulated pattern (Figure 4a) thus establishing the bulk phase purity of IPM_325p1.

We observed a considerable shift in the PXRD pattern of the asynthesized compound upon exposure to air. This shift may be attributed to the sliding of the interpenetrated networks at ambient conditions due to escaping of low boiling guests from the voids. The PXRD profile showed an entirely new pattern thus confirming the formation of another structure. Another interesting observation from the PXRD pattern is that the intensity profile at $2\theta = 5.37°$ peak exhibited a highly sharp peak after long exposure to ambient conditions. From this it can be inferred that IPM-325_p2 retains a crystalline nature after exposure to ambient conditions thus establishing the air stability of this phase. Due to the drastic shift observed in PXRD, we sought to obtain a single crystal of IPM-325_p2 after air exposure but due to the poor quality of the crystal we were unable to obtain single crystal data. Thermogravimetric analysis of the IPM-325_p2 present at ambient conditions showed ~10% weight loss around ~103 °C owing to the presence of occluded guests molecules present within the pores with overall loss of 20% observed until 220 °C. (Figure 4b). To obtain the desolvated phase of compound the material was heated at elevated temperature (160 °C) overnight with vacuum and further characterized via both PXRD and TGA measurements. Desolvated phase was characterized using thermogravimetric analysis which showed no initial loss in the desolvated phase of compound ~220 °C after which a gradual loss was observed which can be correlated to the decomposition of the framework. Powder X-ray diffraction measurement of the desolvated phase exhibited an entirely new pattern with the retention of original peak at $2\theta = 5.37°$ along with the emergence of some new peaks (Figure 4a). Thus, IPM-325_p2 showed another structural transformation to yield the desolvated phase IPM-325_p3 (p represents phase). Also, the peak at $2\theta = 5.37°$ maintained its intensity thus from which we can infer that the desolvated framework is highly crystalline in nature. Low temperature N_2 adsorption was performed, which revealed slow opening of the pore (Figure 5).

Thus, we have successfully synthesized and characterized a bi-porous MOF; IPM-325_p1 wherein SiF_6^{2-} anion acts as an inorganic pillar. IPM-325_p2 shows structural transformation as analyzed by PXRD measurements. Further investigations regarding the structural aspect of the desolvated phase IPM-325_p3 and its sorption measurements are currently being pursued in our lab.

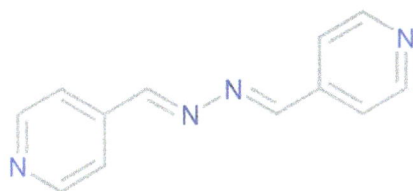

Scheme 1. Chemical diagram of linker L = ((1*E*,2*E*)-1,2-bis(pyridin-4-ylmethylene)hydrazine).

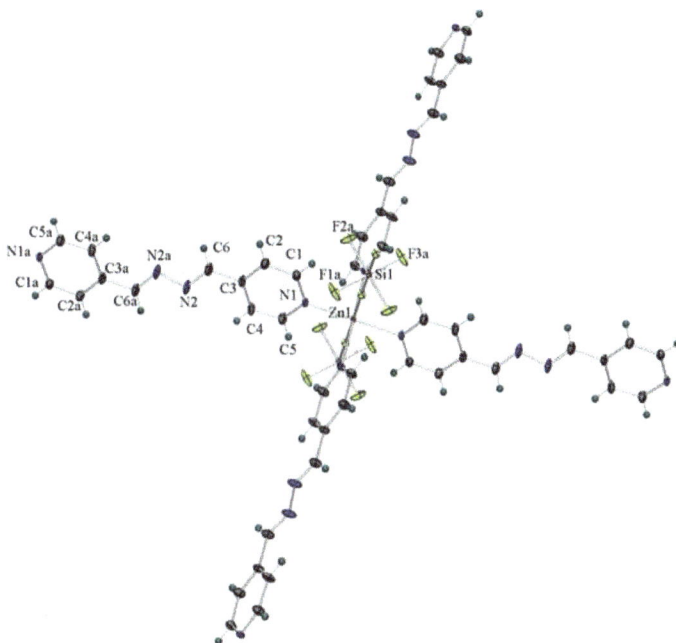

Figure 1. Ortep diagram showing coordination sphere around Zn(II) metal ion in thermal ellipsoids (50% probability).

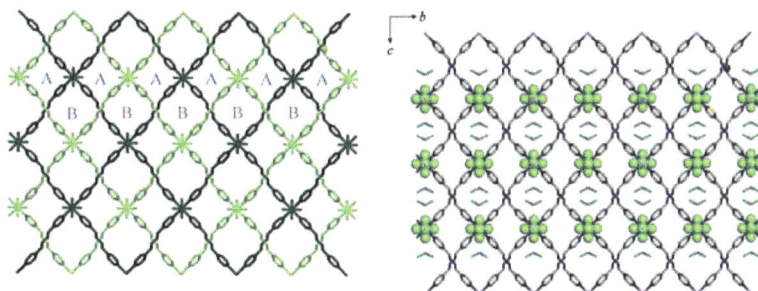

Figure 2. (**a**) Packing diagram along *a*-axis of IPM-325_p1 with guests solvent molecules present within the one-dimensional pore channel; (**b**) Packing diagram of IPM-325_p1 along *a*-axis depicting two-fold interpenetration along with two kinds of pore. (Hydrogen atoms have been omitted for clarity) Color: Grey—C, Light Blue—N, Dark Blue—Zn, Green—Fluorine, Red—Silicon, Light green—Chlorine. Solvent hidden for clarity.

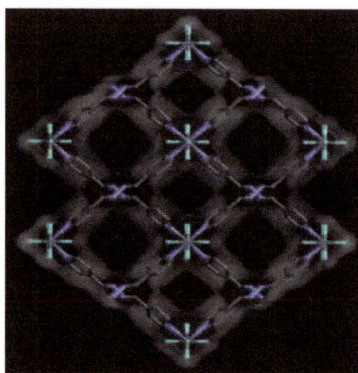

Figure 3. Packing diagram showing potential voids of framework. Color: Grey—C, Light Blue—N, Blue—Zn, Light green—Chlorine.

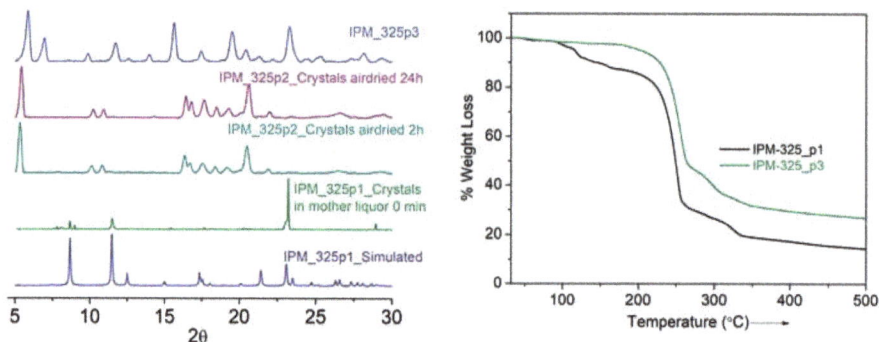

Figure 4. (a) Powder X-Ray diffraction patterns of simulated (IPM-325_p1) Color Code—Blue, IPM-325p1_crystals soaked in mother liquor 0 min Color Code—Green, Air exposed after 2 h (IPM-325_p2) Color Code—Cyan, Air exposed after 24 h (IPM-325_p2) Color Code—Purple, Desolvated framework IPM_325_p3; (b) Thermogravimetric analysis of IPM-325_p1 and IPM-325_p3.

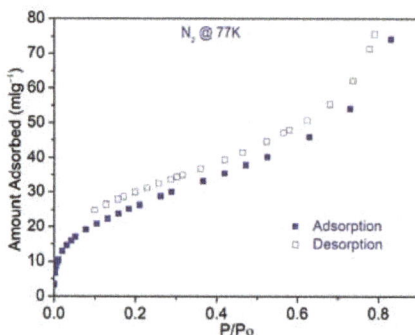

Figure 5. Low temperature nitrogen gas isotherm for IPM_325p3 at 77 K.

Table 1. Crystal data for IPM-325_p1.

Crystal Data	IPM_MOF-325-p1
Formula Crystal	{[Zn(L)$_2$ (SiF$_6$)]·(CH$_2$Cl$_2$)·xG}$_n$ Monoclinic
CCDC	1,586,047
a/Å	7.73 (9)
b/Å	20.36 (3)
c/Å	12.09 (14)
α/°	90
β/°	102.630 (2)
γ/°	90
Cell volume/Å^3	1857.6 (4)
Space group	C 2/m
Z	2
Dx/Mg m^{-3}	1.426
R	5.4

Acknowledgments: B.M. is thankful to CSIR for research funding. S.S. is thankful to IISER-Pune for research fellowship. We thank Aamod V. Desai for his inputs. We acknowledge IISER-Pune for research facilities. We would also like to acknowledge SERB (Project No. EMR/2016/000410) for generous financial support and DST-FIST (SR/FST/CSII-023/2012) for Single Crystal instrument facility.

Author Contributions: S.K.G. supervised the project. B.M. and S.S. carried out the experimentation and wrote the manuscript.

Conflicts of Interest: The authors declare no conflict of interest.

References

1. Furukawa, H.; Cordova, K.E.; O'Keeffe, M.; Yaghi, O.M. The chemistry and Applications of metal-organic frameworks. *Science* **2013**, *341*, 974. [CrossRef] [PubMed]

2. Ouay, B.L.; Kitagawa, S.; Uemura, T. Opening of an accessible microporosity in an otherwise nonporous metal–organic framework by polymeric guests. *J. Am. Chem. Soc.* **2017**, *139*, 7886–7892. [CrossRef] [PubMed]

3. Czaja, A.U.; Trukhan, N.; Muller, U. Industrial applications of metal–organic frameworks. *Chem. Soc. Rev.* **2009**, *38*, 1284–1293. [CrossRef] [PubMed]

4. Lustig, W.P.; Mukherjee, S.; Rudd, N.D.; Desai, A.V.; Li, J.; Ghosh, S.K. Metal–organic frameworks: Functional luminescent and photonic materials for sensing applications. *Chem. Soc. Rev.* **2017**, *46*, 3242–3285. [CrossRef] [PubMed]

5. Lee, J.; Farha, O.K.; Roberts, J.; Scheidt, K.A.; Nguyen, S.T.; Hupp, J.T. Metal–organic framework as catalysts. *Chem. Soc. Rev.* **2009**, *38*, 1450. [CrossRef] [PubMed]

6. Biswas, S.; Stock, N. Synthesis of metal–organic frameworks (MOFs): Routes to various MOF topologies, morphologies, and composites. *Chem. Rev.* **2012**, *112*, 933–969.

7. Zhou, H.-C.J.; Kitagawa, S. Metal–Organic Frameworks (MOFs). *Chem. Soc. Rev.* **2014**, *43*, 5415–5418. [CrossRef] [PubMed]

8. Lu, W.; Wei, Z.; Gu, Z.-Y.; Liu, T.-F.; Park, J.; Park, J.; Tian, J.; Zhang, M.; Zhang, Q.; Gentle, T., III; et al. Tuning the structure and function of metal–organic frameworks via linker design. *Chem. Soc. Rev.* **2014**, *43*, 5561–5593. [CrossRef] [PubMed]

9. Lalonde, M.; Bury, W.; Karagiaridi, O.; Brown, Z.; Hupp, J.T.; Farha, O.K. Transmetalation: Routes to metal exchange within metal–organic frameworks. *J. Mater. Chem. A* **2013**, *1*, 5453–5468. [CrossRef]

10. Brozek, C.K.; Dinca, M. Cation exchange at the secondary building units of metal–organic frameworks. *Chem. Soc. Rev.* **2014**, *43*, 5456–5467. [CrossRef] [PubMed]

11. Schoedel, A.; Li, M.; Li, D.; O'Keeffe, M.; Yaghi, O.M. Structures of metal–organic frameworks with rod secondary building units. *Chem. Rev.* **2016**, *116*, 12466–12535. [CrossRef] [PubMed]

12. Rimoldi, M.; Howarth, A.J.; DeStefano, M.R.; Lin, L.; Goswami, S.; Li, P.; Hupp, J.T.; Farha, O.K. Catalytic zirconium/hafnium-based metal–organic frameworks. *ACS Catal.* **2017**, *7*, 997–1014. [CrossRef]

13. Noro, S.-I.; Kitaura, R.; Kondo, M.; Kitagawa, S.; Ishii, T.; Matsuzaka, H.; Yamashita, M. Framework engineering by anions and porous functionalities of Cu(II)/4,4′-bpy coordination polymers. *J. Am. Chem. Soc.* **2002**, *124*, 2568–2583. [CrossRef] [PubMed]

14. Karmakar, A.; Desai, A.V.; Ghosh, S.K. Ionic metal-organic frameworks (iMOFs): Design principles and applications. *Coord. Chem. Rev.* **2016**, *307*, 313–341. [CrossRef]

15. Manna, B.; Desai, A.V.; Ghosh, S.K. Neutral N-donor ligand based flexible metal-organic frameworks. *Dalton Trans.* **2016**, *45*, 4060–4072. [CrossRef] [PubMed]

16. Eddaoudi, M.; Sava, D.F.; Eubank, J.F.; Adil, K.; Guillerm, V. Zeolite-like metal-organic frameworks (ZMOFs): Design, synthesis, and properties. *Chem. Soc. Rev.* **2015**, *44*, 228–249. [CrossRef] [PubMed]

17. Subramaniam, S.; Zaworotko, M.J. Porous solids by design: [Zn(4,4'-bpy)$_2$(SiF$_6$)]$_n$ x DMF, a single framework octahedral coordination polymer with large square channels. *Angew. Chem. Int. Ed.* **1995**, *34*, 2127. [CrossRef]

18. Noro, S.-I.; Kitagawa, S.; Kondo, M.; Kenji, S. New methane adsorbent, porous coordination polymer, [Cu(SiF$_6$)(4,4'-bipyridine)$_2$]$_n$. *Angew. Chem. Int. Ed.* **2000**, *39*, 2081–2084. [CrossRef]

19. Shekhah, O.; Belmabkhout, Y.; Chen, Z.; Guillerm, V.; Cairns, A.; Adil, K.; Eddaoudi, M. Made-to-order metal–organic frameworks for trace carbon dioxide removal and air capture. *Nat. Commun.* **2014**, *5*, 4228. [CrossRef] [PubMed]

20. Burd, S.D.; Ma, S.; Perman, J.A.; Sikora, B.J.; Snurr, R.Q.; Thallapally, P.K.; Tian, J.; Wojtas, L.; Zaworotko, M.J. Highly Selective Carbon Dioxide Uptake by [Cu(bpy-n)$_2$(SiF$_6$)] (bpy-1 = 4,4' -Bipyridine; bpy-2 = 1,2-Bis(4-pyridyl)ethane). *J. Am. Chem. Soc.* **2012**, *134*, 3663. [CrossRef] [PubMed]

21. Nugent, P.; Belmabkhout, Y.; Burd, S.D.; Cairns, A.J.; Luebke, R.; Forrest, K.; Pham, T.; Ma, S.; Space, B.; Wojtas, L.; et al. Porous materials with optimal adsorption thermodynamics and kinetics for CO$_2$ separation. *Nature* **2013**, *495*, 80–84. [CrossRef] [PubMed]

22. O'Nolan, D.; Kumar, A.; Zaworotko, M.J. Water Vapour Sorption in Hybrid Pillared Square Grid Materials. *J. Am. Chem. Soc.* **2017**, *139*, 8508–8513. [CrossRef] [PubMed]

23. Zhang, Z.; Yang, Q.; Cui, X.; Yang, L.; Bao, Z.; Ren, Q.; Xing, H. Sorting of C4 olefins with interpenetrated hybrid ultramicroporous materials by combining molecular recognition and size-sieving. *Angew. Chem. Int. Ed.* **2017**, *56*, 16282–16287. [CrossRef] [PubMed]

24. Schneemann, A.; Bon, V.; Schwedler, I.; Senkovska, I.; Kaskel, S.; Fischer, R.A. Flexible metal–organic frameworks. *Chem. Soc. Rev.* **2014**, *43*, 6062–6096. [CrossRef] [PubMed]

25. Chang, Z.; Yang, D.-H.; Xu, J.; Hu, T.-L.; Bu, X.-H. Flexible metal–organic frameworks: Recent advances and potential applications. *Adv. Mater.* **2015**, *27*, 5432–5441. [CrossRef] [PubMed]

26. Horike, S.; Shimomura, S.; Kitagawa, S. Soft porous crystals. *Nat. Chem.* **2009**, *1*, 695–704. [CrossRef] [PubMed]

27. Serre, C.; Millange, f.; Thouvenot, C.; Nogues, M.; Marsolier, G.; Louer, D.; Ferey, G. Very large breathing effect in the first nanoporous chromium(iii)-based solids: MIL-53 or Cr^{III}(OH)·{O$_2$C−C$_6$H$_4$−CO$_2$} ·{HO$_2$C−C$_6$H$_4$−CO$_2$H}$_x$·H$_2$O$_y$. *J. Am. Chem. Soc.* **2002**, *124*, 13519–13526. [CrossRef] [PubMed]

28. Hamon, L.; Llewellyn, P.L.; Devic, T.; Ghoufi, A.; Clet, G.; Guillerm, V.; Pirngruber, G.D.; Maurin, G.; Serre, S.; Driver, G.; et al. Co-adsorption and separation of CO$_2$-CH$_4$ mixtures in the highly flexible MIL-53(Cr) MOF. *J. Am. Chem. Soc.* **2009**, *131*, 17490–17499. [CrossRef] [PubMed]

29. *SAINT Plus*, Version 7.03; Bruker AXS Inc.: Madison, WI, USA, 2004.

30. Krause, L.; Herbst-Irmer, R.; Sheldrick, G.M.; Stalke, D. Comparison of Silver and molybdenum microfocus X-ray sources for single crystal structure determination. *J. Appl. Cryst.* **2015**, *48*, 3–10. [CrossRef] [PubMed]

31. Sheldrick, G.M. A short history of SHELX. *Acta Crystallogr. Sect. A* **2008**, *64*, 112–122. [CrossRef] [PubMed]

32. Sheldrick, G.M. SHELXT—Integrated space-group and crystal-structure determination. *Acta Cryst.* **2015**, *C71*, 3–8. [CrossRef] [PubMed]

33. Farrugia, L. *WinGX*, version 1.80.05; University of Glasgow: Glasgow, Scotland, 2009.

34. Spek, A.L. PLATON SQUEEZE: A tool for the calculation of the disordered solvent contribution to the calculated structure factors. *Acta Cryst.* **2015**, *C71*, 9–18.

35. Bisht, K.K.; Suresh, E. Spontaneous Resolution to Absolute Chiral Induction: Pseudo-Kagome Type Homochiral Zn(II)/Co(II) Coordination Polymers with Achiral Precursors. *J. Am. Chem. Soc.* **2013**, *135*, 15690.

crystals

MDPI

Article

A 12-Fold ThSi$_2$ Interpenetrated Network Utilizing a Glycine-Based Pseudopeptidic Ligand

Edward Loukopoulos, Alexandra Michail and George E. Kostakis *

Department of Chemistry, School of Life Sciences, University of Sussex, Brighton BN1 9QJ, UK;
E.Loukopoulos@sussex.ac.uk (E.L.); alexandra_michail@hotmail.com (A.M.)
* Correspondence: G.Kostakis@sussex.ac.uk; Tel.: +44-1273-877-339

Received: 21 November 2017; Accepted: 15 January 2018; Published: 18 January 2018

Abstract: We report the synthesis and characterization of a 3D Cu(II) coordination polymer, [Cu$_3$(L^1)$_2$(H$_2$O)$_8$]·8H$_2$O (**1**), with the use of a glycine-based tripodal pseudopeptidic ligand (H$_3$L^1 = N,N',N''-tris(carboxymethyl)-1,3,5-benzenetricarboxamide or trimesoyl-tris-glycine). This compound presents the first example of a 12-fold interpenetrated ThSi$_2$ (**ths**) net. We attempt to justify the unique topology of **1** through a systematic comparison of the synthetic parameters in all reported structures with H$_3$L^1 and similar tripodal pseudopeptidic ligands. We additionally explore the catalytic potential of **1** in the A^3 coupling reaction for the synthesis of propargylamines. The compound acts as a very good heterogeneous catalyst with yields up to 99% and loadings as low as 3 mol %.

Keywords: coordination polymer; copper; interpenetration; 12-fold; catalysis; A^3 coupling

1. Introduction

Ever since their popularization in the last decade, coordination polymers (CPs), also known as metal–organic frameworks (MOFs), have become one of the most prominent branches of inorganic and materials chemistry, in no small part due to the extensive variety of their applications [1–8]. A key factor for the rise of CPs as functional materials has been the development of rational synthetic routes towards the optimization of their application potential [9–12]. As a result, the strategic selection of a suitable ligand for the synthesis of CPs during the design of a new system is critical.

In recent years, a surge towards biologically derived CPs has been observed [13]. This type of compounds is typically constructed from bioligands such as amino acids, peptides, or nucleobases, which can offer a large variety for exploitation regarding possible coordination sites, the presence of functional groups, the degree of flexibility, and the potential formation of strong and weak interactions. This plethora of options has led to multiple reports of biologically related CPs with interesting applications: for example, porous CPs with adenine- [14], serine- [15], and dipeptide-based [16] ligands have been used successfully for the selective capture of CO$_2$. Furthermore, several ligands with amino acids or peptides have been employed with transition metals to induce chirality in various catalytic procedures [17–19]. Finally, CPs containing adenine have also been reported to have potential sensing [20] and drug storage [21] capabilities.

Pseudopeptidic ligands are another interesting type of biologically related linkers. In this case, a scaffold, which may be aromatic or nonaromatic, has amino acids or oligopeptides attached to it (Scheme 1). This type of ligands can offer multiple positions for coordination, as well as unlimited choices in flexibility, aromaticity, and hydrogen bond formation.

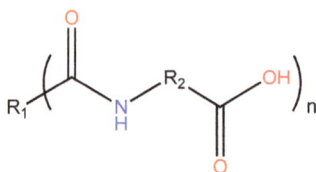

Scheme 1. General motif of pseudopeptidic ligands.

Our group [22–27] and others [28,29] have focused specifically on the coordination chemistry of pseudopeptidic ligands which retain a rigid aromatic scaffold, and introduced to it flexible amino acids. This strategy has produced a great variety of coordination polymers with many applications, such as luminescence [30], reversible water loss [27,30,31], magnetism [29], homochirality [18], and catalysis [26]. Additionally, this type of pseudopeptidic ligands are of great interest from a crystal engineering point of view and can lead to fascinating topologies due to: (a) the varying nature of the aromatic scaffold; (b) the possible formation of strong and weak hydrogen bonds as well as pi interactions; (c) the length of the amino acids; (d) the coordination abilities of the metal ion used.

Our most recent work in this project studied the influence of the Cu(II) salt on the resulting CPs when the pseudopeptidic ligand isophthaloylbis-β-alanine (H_2IBbA) was used [25]. Encouraged by the obtained result, we opted to investigate this effect on additional pseudopeptidic ligands. Having also in mind (a) our ongoing interest on CPs and their catalytic properties [32–34] and (b) the commonly limited solubility of these bulky pseudopeptidic CPs in organic solvents, we theorized that the resulting compounds could have an interesting activity as heterogeneous catalysts. Therefore, we herein present the synthesis, characterization, and topological evaluation of a novel compound formulated as $[Cu_3(L^1)_2(H_2O)_8] \cdot 8H_2O$ (**1**), where H_3L^1 is N,N',N''-tris(carboxymethyl)-1,3,5-benzenetricarboxamide [35], also commonly known as trimesoyl-tris-glycine (Scheme 2). Furthermore, we report the catalytic activity of **1** in the metal-catalyzed multicomponent reaction (MCR) of an aldehyde, an amine, and an alkyne, also known as the A^3 coupling, towards the synthesis of propargylamines.

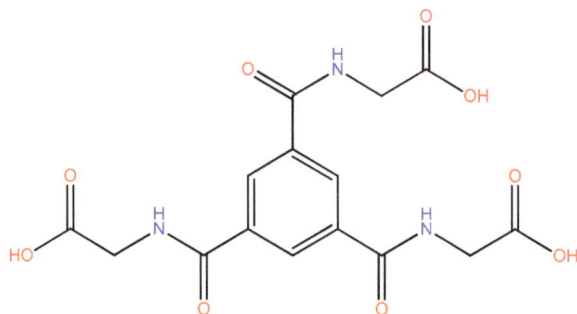

Scheme 2. The organic ligand H_3L^1 used in this study.

2. Experimental Section

2.1. Materials

Chemicals (reagent grade) were purchased from Sigma Aldrich (Gillingham, UK), Acros Organics (Loughborough, UK), and Alfa Aesar (Lancashire, UK). All experiments were performed under aerobic conditions using materials and solvents as received. Ligand H_3L^1 was synthesized according to the reported procedure [23].

2.2. Instrumentation

IR spectra of the samples were recorded over the range of 4000–650 cm^{-1} on a Perkin Elmer Spectrum One FT-IR spectrometer (Seer Green, UK) fitted with a UATR polarization accessory. EI-MS was performed on a VG Autospec Fissions instrument (EI at 70 eV, SIS, Ringoes, NJ, USA). TGA analysis was performed on a TA Instruments Q-50 model (TA, Surrey, UK) under nitrogen and at a scan rate of 10 °C/min. NMR spectra were measured on a Varian VNMRS solution-state spectrometer (Agilent, Stockport, UK) at 30 °C. Chemical shifts were quoted in parts per million (ppm). Coupling constants (J) were recorded in Hertz (Hz).

2.3. X-ray Crystallography

Data for compound **1** were collected (ω-scans) at the University of Sussex using an Agilent Xcalibur Eos Gemini Ultra diffractometer (Agilent, Stockport, UK) with a CCD plate detector (Agilent, Stockport, UK) under a flow of nitrogen gas at 173(2) K and Mo Kα radiation (λ = 0.71073 Å). CRYSALIS CCD and RED software (version 1.171.38.41) were used, respectively, for data collection and processing. Reflection intensities were corrected for absorption by the multiscan method. The structure was determined using Olex2 [36], solved using SHELXT (version 2015) [37,38], and refined with SHELXL-2014 [39]. All non-H atoms were refined with anisotropic thermal parameters, and H-atoms were introduced at calculated positions and allowed to ride on their carrier atoms. For certain water lattice molecules, the introduction of H atoms at calculated positions led to an unsatisfactory structure solution with certain short intramolecular D–H···H–D distances. Because of the poor quality of the crystallographic data in three different datasets, we were unable to locate these H atoms manually. For this reason, bond distances and angles regarding these atoms were not mentioned further. Additional measurements of elemental and TGA analysis were performed to conclusively validate the suggested formula and structure. Crystal data and structure refinement parameters for **1** are given in Table S1. The geometric and crystallographic calculations were performed using PLATON (version 1.18) [40], Olex2 (version 1.2) [36], and WINGX [41] packages (version 2014.1). The graphics were prepared with Crystal Maker and MERCURY [42], CCDC 1586912.

2.4. Synthetic Procedures

2.4.1. Synthesis of [Cu$_3$(L^1)$_2$(H$_2$O)$_8$]·8H$_2$O (**1**)

Method 1: 0.5 mmol (0.190 g) of H$_3$L^1 and 0.75 mmol (110 µL) of Et$_3$N were dissolved in 15 mL of H$_2$O while stirring to produce a colourless solution. To this, 1 mmol (0.232 g) of Cu(NO$_3$)$_2$·2.5H$_2$O was added. The resulting light blue solution was stirred for a further 20 min at room temperature. After the end of the reaction, the solution was filtrated and then carefully layered over Me$_2$CO at a respective ratio of 1:2. After 6 days, large blue block crystals of **1**, as well as colourless needles of H$_3$L^1 were formed. Yield: 27% (based on Cu). *Method 2*: the same procedure as above was followed, but a solution of H$_2$O/MeOH (10:1) was used instead of H$_2$O to produce only crystals of **1**. Yield: 18% (based on Cu). Selected IR peaks (cm^{-1}): 3253 (br), 1626 (m), 1557 (m), 1499 (w), 1434 (m), 1403 (m), 1304 (w), 1003 (w), 909 (w), 733 (m). Elemental analysis for C$_{30}$H$_{62}$Cu$_3$N$_6$O$_{34}$: calcd. C 43.89, H 5.04, N 6.78; found C 43.84, H 5.06, N 6.89.

2.4.2. General Catalytic Protocol for A^3 Coupling

A mixture of aldehyde (0.5 mmol), amine (0.55 mmol), alkyne (0.6 mmol), Cu catalyst **1** (3 mol%, based on aldehyde amount), and 2-propanol (2 ml) was added into a sealed tube and stirred at 90 °C for 12 h. After completion of the reaction, the mixture was allowed to cool down. The slurry was then filtered to withhold the catalyst, and the filtrate was evaporated under vacuum. The yield of the propargylamine products were then determined by their ^{1}HNMR spectra, which were compared with the data reported in the corresponding literature [34].

3. Results

3.1. Crystal Structure Description

Compound **1** crystallizes in the noncentrosymmetrical monoclinic *Pn* space group. X-ray determination of the crystal structure reveals the formation of an interpenetrated neutral three-dimensional coordination polymer (Figures 1 and 2). The asymmetric unit of **1** consists of 3 copper centres, 2 fully deprotonated ligand $(L^1)^{3-}$ molecules, and a total of 16 water molecules; out of these, 8 act as terminal ligands and 8 are present in the lattice. In both $(L^1)^{3-}$ molecules, each of the three carboxylate groups coordinate to each of the three Cu centers. Additionally, in both ligands, the three amido groups exist in a *trans* conformation and all of them are in *anti* conformation (Figure 3). Cu1 is coordinated to six atoms and exhibits a distorted octahedral geometry (s/h = 1.04, ϕ = 52.9° [43]). The equatorial positions of this octahedron are occupied by four carboxylate oxygen atoms deriving from two different ligands. Cu2 and Cu3 are each coordinated to five atoms and exhibit a square pyramidal geometry (τ = 0.06 for Cu2, 0.17 for Cu3 [44]). In the coordination environment of both metal centers, the basal plane consists of two oxygen atoms from two different ligand molecules, as well as two oxygen atoms from terminal water molecules. An oxygen atom from another terminal water molecule occupies the apical position in both cases. Selected bond lengths are listed in Table 1. The Cu–Cu distances between the metal centers range from 11.400(3) to 14.176(3) Å. Furthermore, the crystal structure of **1** is stabilized by strong intermolecular O–H···O hydrogen bonds, which involve the oxygen atoms of all 16 water molecules as donors. The atoms involved as acceptors in these bonds are oxygen atoms of either water molecules or the carbonyl group of the amide.

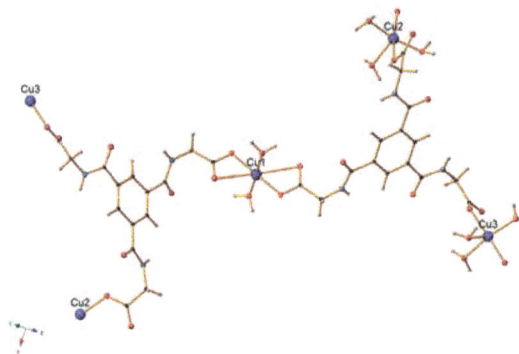

Figure 1. The crystal structure of **1**. Lattice water molecules are omitted for clarity. Colour code: Cu (blue), C (black), H (light pink), N (light blue), O (red).

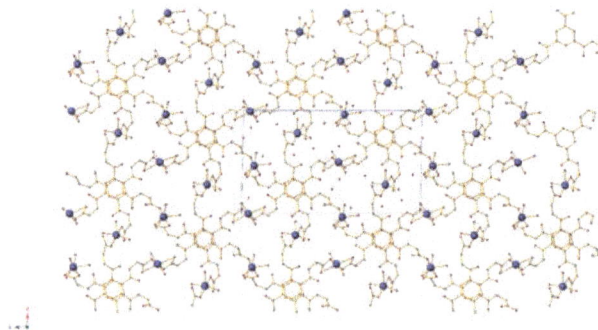

Figure 2. Packing diagram for **1** along the *a0c* plane.

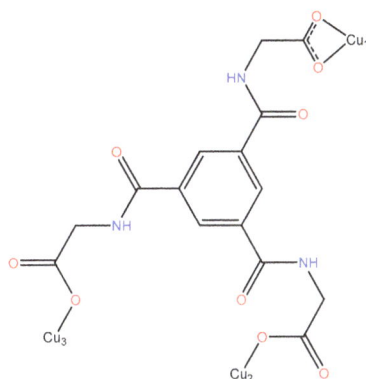

Figure 3. Coordination mode of the $(L^1)^{3-}$ organic ligand found in **1**.

Table 1. Selected bond lengths (Å) for **1**.

Bond	Å
Cu1–O1	1.932(11)
Cu1–O2	2.668(11)
Cu1–O10	1.964(10)
Cu1–O11	2.611(11)
Cu1–O19	1.938(11)
Cu1–O20	1.931(10)
Cu2–O8	1.942(11)
Cu2–O28	1.991(9)
Cu2–O29	2.191(11)
Cu2–O30	1.959(10)
Cu2–O14 [1]	1.927(11)
Cu3–O5	1.928(10)
Cu3–O22	1.925(12)
Cu3–O23	2.005(11)
Cu3–O24	2.343(11)
Cu3–O18 [2]	1.943(10)
Cu1–Cu2	11.400(3)
Cu1–Cu3	14.176(3)
Cu2–Cu3	12.554(3)

Symmetry Operators: [1] $-3/2 + X, 2 - Y, 1/2 + Z$; [2] $+X, -2 + Y, 1 + Z$.

3.2. Topological Analysis

The complicated structure of compound **1** can be simplified into a net considering each ligand as a three-connected node and each metal center as a two-connected node; therefore, the two-connected nodes are not further considered for the classification. The final outcome of the topological analysis of the three-dimensional coordination polymer **1**, with the use of the TOPOS software [45] and the standard representation methodology, is a three-connected, 12-fold interpenetrated symmetric **ths** net (Figure 4). According to a literature survey in the TOPOS and CCDC databases, compounds EJISAS [46] and KOBFEN [47] can be also represented as 12-fold **ths** nets; however, this simplification derives when a standard representation is selected. The topological analysis of the latter two compounds using the standard cluster representation and considering the $Cu(O_2)Cu$ as nodes [48], yields a 12-fold interpenetrated diamond (**dia**) net, therefore compound **1** represents the first example of a standard 12-fold interpenetrated **ths** net.

(a)

(b)

(c)

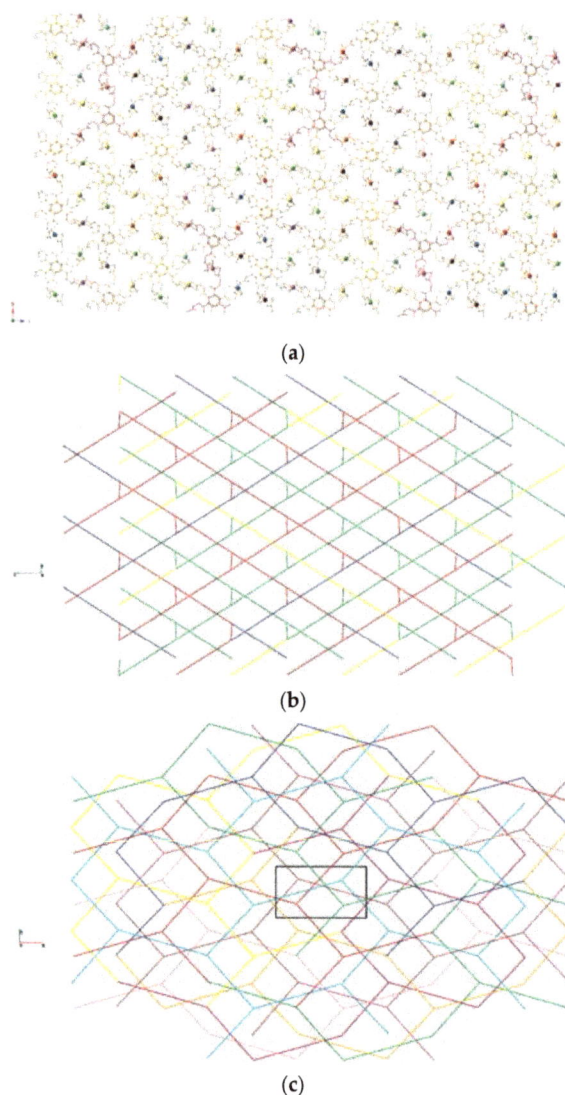

Figure 4. Simplified coloured versions of the 3D 12-fold interpenetrated **ths** network found in **1** along the *a0c* (**a**), *b0c* (**b**), *b0a* (**c**) planes.

3.3. TGA and IR Studies

To examine the thermal behavior and stability of **1**, TGA was carried out between room temperature and 800 °C under N_2 atmosphere. This analysis (Figure S1) shows that the first mass loss is continuous, as it begins in the region of 50°C and is completed at approximately 150 °C. This is attributed to the loss of eight lattice and eight ligated water molecules, in good agreement with the theoretical value (calc.: 23.60%, theor.: 23.25%). The remaining framework is then relatively stable up to ~310 °C, where it is subjected to a further mass loss due to decomposition to CuO (calc.: 63.14%, theor.: 65.14%). The reported peaks in the IR spectrum of **1** (Figure S2) are in good agreement with the crystallographic data. A broad absorption peak is found at 3253 cm^{-1} and is attributed to the stretching vibration of the O–H bonds. The peak at 1622 cm^{-1} is due to the presence of the noncoordinated carbonyl group of the amide, in good agreement with previously reported values for similar [23] compounds. Furthermore, peaks at 1557 and 1403 cm^{-1} can be attributed to the $\nu(CO_2)_{as}$ and symmetric $\nu(CO_2)_s$ bands of coordinated carboxylate groups, respectively. Finally, some peaks related possibly to C–H bending vibrations appear at 909 and 733 cm^{-1}.

3.4. Synthetic Aspects

Our initial efforts for the synthesis of **1** involved experiments in various ratios of water/alcohol media, based on our previous experiences with H_3L^1 [23] as well as the related literature [30]. However, no crystals were obtained in this case. The protocol was therefore modified with various techniques and ratios in order to facilitate crystallization. After extensive screening, liquid diffusion in acetone was found to be the only effective technique amongst the tested ones. The use of other suitable secondary crystallization solvents (e.g., acetonitrile) led, instead, to amorphous material. It is worth noting that the water/alcohol mix seems to be critical for the pure synthesis of **1**, as a similar experiment in H_2O also yielded crystals of the organic ligand; however, no MeOH molecules were found in the structure, despite their potential participation in H-bonding.

While the topology of **1** has not been observed before, the afforded compound is not the only structure which contains a Cu(II) source and the H_3L^1 ligand. In fact, a search in the CCDC [49] revealed a variety of structures, but all of these show a different topology. To shed more light into this as well as attempt to rationalize the synthesis, we opted to perform a more systematic search in the literature for similar tripodal pseudopeptidic ligands. This narrowed our results to a total of 28 reported coordination compounds, with 3 different ligands depending on the varying amino acid: either Glycine (H_3L^1), L-Alanine (H_3L^2), or D-Alanine (H_3L^3) (Scheme 3). To provide a full insight, we included a full list of factors that could point towards the resulting differences. These parameters included the metal ion, the synthetic conditions including solvent and temperature, and the presence of a base or a second organic linker. These are listed in detail in Table 2.

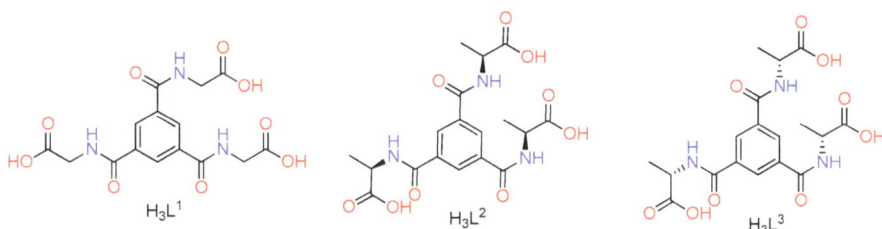

Scheme 3. The tripodal pseudopeptidic ligands compared in this study.

In regard to the H_3L^1 glycine-based ligand, a comparison between our result (Entry 1) and the rest of the reported Cu(II) compounds (entries 2–5) already revealed major influences of these parameters. Compound **1** was synthesized using $Cu(NO_3)_2 \cdot 2.5H_2O$, while, in the rest of the relevant entries, $CuCl_2 \cdot 2H_2O$ was used as the metal source. The role of the metal ion in the resulting structure has already been reported, especially for Cu(II) sources in similar pseudopeptidic ligands. Therefore, our result further confirmed this effect. The rest of the parameters revealed additional interesting information: a comparison of entries 2–4 showed that the presence and amount of base (and as a consequence, the tuning of pH) led to different structures; in the case of entry 4, the base (pyridine) actually coordinated to the metal center, which led to a 2D coordination polymer instead of a 3D, and to a less exciting topology. Regarding the synthetic conditions between these entries, a possible temperature effect over time could be observed. Efforts to obtain a crystal structure using $CuCl_2 \cdot 2H_2O$ and the synthetic method of **1**, or $Cu(NO_3)_2 \cdot 2.5H_2O$ and solvothermal conditions were unfortunately unsuccessful. However, we could obtain a wider scope for conclusions by bringing also the glycine-based compounds with other metals (entries 6–20) into the comparison. Through this, it is worth noting the following: (a) the compounds of entries 2 and 6–9 had a general formula of $[M(L^1)(H_2O)_3]_2[M(H_2O)_6]\cdot(H_2O)_3$ regardless of the synthetic method; (b) our attempts to utilize our synthetic method with other metals (Co, Zn, Mn) resulted in the same crystal structures in entries 2 and 6–9, conclusively proving that the synthetic procedures were not the prevalent factor in order to get structures with the 12-fold topology; (c) as expected, the presence of a second organic linker led to even more unpredictable structures. Interestingly, a comparison between entries 5 and 12 (Cu- and Co-based respectively), in which the same linker (bpp) and similar synthetic methods were employed, revealed significant differences in the resulting products, further pointing to the lesser importance of the conditions compared to the choice of metal; (d) a comparison between Ca(II)-based compounds $[Ca_6(L^1)_4(H_2O)_{14}](H_2O)_3$ and $[Ca_2(HL^1)_2(\mu\text{-}H_2O)(H_2O)_5]\cdot3H_2O$ (entries 18 and 19 respectively), which were synthesized under very similar methods but with a different Ca(II) source (chloride for entry 18, nitrate for 19), further pointed towards the metal ion influence; (e) only the metals with flexibility in their coordination environment and geometry (copper, alkaline earth metals, lanthanides) provided any cases of structural variety. Interestingly, the largest variety of compounds was observed when Cu(II) sources were employed.

Table 2. Overview of the synthetic parameters and topology of all reported compounds with tripodal pseudopeptidic ligands.

Entry	Metal Salt	L	Additive [a]	Conditions	Formula	Ref.
1	$Cu(NO_3)_2\cdot2.5H_2O$	H_3L^1	Et_3N	$rt/H_2O/MeOH$ (10:1)/Me_2CO	$[Cu_3(L^1)_2(H_2O)_8]\cdot8H_2O$	[b]
2	$CuCl_2\cdot2H_2O$	H_3L^1	None	100 °C/3 h/H_2O/DMF (1:2)	$[Cu(L^1)(H_2O)_3]_2[Cu(H_2O)_6]\cdot(H_2O)_3$	[50]
3	$CuCl_2\cdot2H_2O$	H_3L^1	py [c]	100 °C/24 h/H_2O/DMF (1:1)	$[Cu_3(L^1)_2(H_2O)_3]\cdot2H_2O$	[50]
4	$CuCl_2\cdot2H_2O$	H_3L^1	py	90 °C/40 h/H_2O/DMF (1:1)	$[Cu_2(L^1)(Py)_2(\mu_3\text{-}OH)]\cdot(H_2O)_2$	[50]
5	$CuCl_2\cdot2H_2O$	H_3L^1	bpp [d]	100 °C/48 h/H_2O/MeOH (1:1)	$[Cu_2(L^1)(bpp)(\mu_3\text{-}OH)]\cdot6H_2O$	[50]
6	$Zn(NO_3)_2\cdot2.5H_2O$	H_3L^1	None	100 °C/48 h/H_2O/MeOH (10:1)	$[Zn(L^1)(H_2O)_3]_2[Zn(H_2O)_6]\cdot(H_2O)_3$	[30]
7	$Ni(NO_3)_2\cdot2.5H_2O$	H_3L^1	None	100 °C/48 h/H_2O/MeOH (10:1)	$[Ni(L^1)(H_2O)_3]_2[Ni(H_2O)_6]\cdot(H_2O)_3$	[30]
8	$Mn(OAc)_2\cdot4H_2O$	H_3L^1	None	100 °C/48 h/H_2O/MeOH (10:1)	$[Mn(L^1)(H_2O)_3]_2[Mn(H_2O)_6]\cdot(H_2O)_3$	[30]
9	$Co(NO_3)_2\cdot6H_2O$	H_3L^1	None	100 °C/48 h/H_2O/MeOH (10:1)	$[Co(L^1)(H_2O)_3]_2[Co(H_2O)_6]\cdot(H_2O)_3$	[30]
10	$Co(NO_3)_2\cdot6H_2O$	H_3L^1	bpy [e]	100 °C/48 h/H_2O	$[Co_{1.5}(L^1)(bpy)_{1.5}(H_2O)_3]\cdot(H_2O)_5$	[51]
11	$Co(NO_3)_2\cdot6H_2O$	H_3L^1	bpe [f]	100 °C/48 h/H_2O	$[Co_{1.5}(L^1)(bpe)_{1.5}(H_2O)_2]$	[51]
12	$Co(NO_3)_2\cdot6H_2O$	H_3L^1	bpp	100 °C/48 h/H_2O	$[Co_2(L^1)(bpp)_2(NO_3)(\mu_2\text{-}H_2O)_2]\cdot(H_2O)_2$	[51]
13	Tb(III)	H_3L^1	None	N/A [g]	$[Tb(L^1)(H_2O)_3]\cdot H_2O$	[31]
14	Gd(III)	H_3L^1	None	N/A	$[Gd(L^1)(H_2O)_3]\cdot H_2O$	[31]
15	Nd(III)	H_3L^1	None	N/A	$[Nd(L^1)(H_2O)_3]\cdot H_2O$	[31]
16	La(III)	H_3L^1	None	N/A	$[La(L^1)(EtOH)(H_2O)_2]\cdot2.5H_2O$	[31]
17	$CaCl_2$	H_3L^1	py	$rt/H_2O/MeOH$ (1:1)	$[Ca(HL^1)(H_2O)_2]$	[52]
18	$CaCl_2$	H_3L^1	py	$rt/H_2O/MeOH$ (1:1)	$[Ca_6(L^1)_4(H_2O)_{14}](H_2O)_3$	[52]
19	$Ca(NO_3)_2\cdot3H_2O$	H_3L^1	NaOAc	$rt/H_2O/EtOH$ (1:1)	$[Ca_2(HL^1)_2(\mu\text{-}H_2O)(H_2O)_5]\cdot3H_2O$	[23]
20	$Sr(NO_3)_2$	H_3L^1	NaOAc	$rt/H_2O/EtOH$ (1:1)	$[Sr_2(HL^1)_2(H_2O)_7]\cdot H_2O$	[23]
21	$Cu(NO_3)_2\cdot3H_2O$	H_3L^2	KOH/am [h]	80 °C/48 h/MeOH/DMF (10:1)	$[Cu_4(HL^2)_2(H_2O)_4(MeO)_4]$	[53]
22	$CuCl_2\cdot2H_2O$	H_3L^2	KOH	$rt/EtOH$/DMF (4:1)	$[Cu_{12}(L^2)_8(H_2O)_{12}]\cdot8EtOH\cdot40H_2O$	[54]
23	$Zn(NO_3)_2\cdot6H_2O$	H_3L^2	bpy/KOH	$rt/H_2O/MeOH$ (3:8)	$[Zn_3(L^2)_2(bpy)_4]\cdot24H_2O$	[55]
24	$Ni(NO_3)_2\cdot2H_2O$	H_3L^2	bpy/KOH	95 °C/48 h/H_2O/EtOH (1:1)	$[Ni_3(L^2)_2(bpy)_4]\cdot2EtOH\cdot14H_2O$	[56]
25	$Co(NO_3)_2\cdot2H_2O$	H_3L^2	bpy/KOH	$rt/H_2O/MeOH$ (3:8)	$[Co_3(L^2)_2(bpy)_4]\cdot28H_2O$	[56]
26	$Cd(NO_3)_2\cdot4H_2O$	H_3L^2	bpy/teda [i]	100 °C/72 h/H_2O/DMF (1:1)	$[Cd_8(L^2)_6(bpy)_3(H_2O)_4](H_3O)_2$	[18]
27	$Ni(NO_3)_2\cdot2H_2O$	H_3L^3	bpy/KOH	95 °C/48 h/H_2O/EtOH (1:1)	$[Ni_3(L^3)_2(bpy)_4]\cdot2EtOH\cdot14H_2O$	[56]
28	$Co(NO_3)_2\cdot2H_2O$	H_3L^3	bpy/KOH	$rt/H_2O/MeOH$ (3:8)	$[Co_3(L^3)_2(bpy)_4]\cdot28H_2O$	[56]
29	$Cd(NO_3)_2\cdot4H_2O$	H_3L^3	bpy/teda	100 °C/72 h/H_2O/DMF (1:1)	$[Cd_8(L^3)_6(bpy)_3(H_2O)_4](H_3O)_2$	[18]

[a] Refers to the use of any base or secondary organic linker. [b] This work. [c] py = pyridine. [d] bpp = (1,3-di(4-pyridyl)propane). [e] bpy = 4,4'-bipyridine. [f] bpe = [1,2-bi(4-pyridyl)ethane]. [g] N/A = Not available. [h] am = ammonium hydroxide, [i] teda = Triethylenediamine.

In regard to the alanine-based compounds (entries 21–29), the presence of an additional methyl group led to completely different compounds and topologies, as expected. However, the stark difference in entries 21 and 22 (in which $Cu(NO_3)_2\cdot3H_2O$ and $CuCl_2\cdot2H_2O$ were employed respectively) once again strongly suggests a metal ion influence towards the resulting product. In summary, when exploring the coordination chemistry of this type of pseudopeptidic ligands, the choice of the metal ion seems to play an important role towards the resulting product and, as a consequence, in the resulting topology and interpenetration.

3.5. Catalytic Studies

The A³ coupling (Scheme 4) has been widely studied in recent years [57–61], as the resulting propargylamines have been proposed as key intermediates in the synthesis of various N-containing biologically active compounds [62–65]. Even though many metal sources and compounds, including CPs [66–69], have been tested as catalysts for this reaction, Cu(II) CPs have been used very rarely [34].

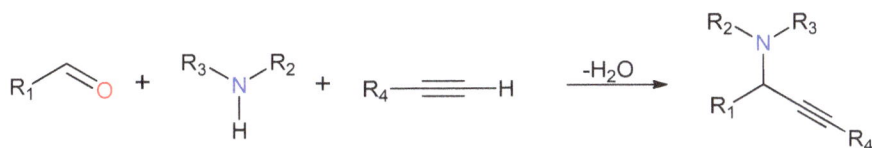

Scheme 4. General overview of the multicomponent reaction of aldehydes, amines, and alkynes (A³ coupling).

In order to test the possible catalytic activity of **1**, initial studies were performed for the A³ coupling of cyclohexane carboxaldehyde, pyrrolidine, and phenylacetylene. After extensive screening, optimal conditions were obtained when the mixture was stirred for 24 h in the presence of 2-propanol (iPrOH) [70], at 90 °C, under air atmosphere, and by adding only 0.03 mmol of compound **1** (in 1 mmol reaction scale of aldehyde). To our delight, these conditions accounted for quantitative yields of the model propargylamine; this accumulated to a turnover number of 33.3 for the catalyst. Additionally, no reaction was observed in the absence of **1**, result that further supports the activity of the catalyst in the studied multicomponent coupling.

We then employed a variety of aldehydes, amines, and alkynes as substrates in order to study the scope of the reaction. Amine screening, as presented in Table 3, entries 1–6, indicated that cyclic secondary amines afford the corresponding propargylamine products in excellent yields, while acyclic secondary amines were found to be slightly less effective. Results of the aldehyde screening (entries 7–10) revealed that aromatic aldehydes show slightly lower reactivity. Furthermore, the reactivity and respective yields were affected by the presence of an electron-donating or electron-withdrawing group in the aldehyde. In comparison, saturated aliphatic aldehydes displayed high reactivity and afforded excellent yields. In regard to the alkyne selection, the employment of either phenylacetylene or 1-hexyne resulted to the corresponding propargylamines in excellent yields when the model aldehyde and amine substrates were also used. The relevant results can be found as entries 1 and 11.

Table 3. Catalytic activity of **1** in the A³ coupling.

Entry	Aldehyde	Amine	Alkyne	Yield [a] (%)
1	cyclohexane carboxaldehyde	pyrrolidine	phenylacetylene	99
2	cyclohexane carboxaldehyde	piperidine	phenylacetylene	99
3	cyclohexane carboxaldehyde	azepane	phenylacetylene	94
4	cyclohexane carboxaldehyde	morpholine	phenylacetylene	99
5	cyclohexane carboxaldehyde	diethylamine	phenylacetylene	77
6	cyclohexane carboxaldehyde	N-methylaniline	phenylacetylene	58
7	benzaldehyde	pyrrolidine	phenylacetylene	67
8	4-methyl benzaldehyde	pyrrolidine	phenylacetylene	61
9	4-methoxy benzaldehyde	pyrrolidine	phenylacetylene	36
10	4-chloro benzaldehyde	pyrrolidine	phenylacetylene	64
11	cyclohexane carboxaldehyde	pyrrolidine	1-hexyne	95
12	cyclohexane carboxaldehyde	pyrrolidine	phenylacetylene	96 [b]

[a] NMR yields based on aldehyde. [b] After the fourth cycle of catalyst use.

The characterization by TGA and IR spectroscopy pointed towards a similar identity of this solid compared to bulk samples of **1** (Figures S3 and S4). The compound showed no solubility in common organic solvents during our tests; therefore, the next step was to study the heterogeneous nature and capabilities of the recycled compound. The catalyst could be easily recovered by filtration after the end of the reactions and then be reused after treatment with acetone and diethyl ether to remove any reagents or product. The simulated and the "as is" synthesized compound powder XRD patterns were in good agreement, however the spectrum of the postcatalysis recovered solid (Figure S5) appeared to be similar to the XRD pattern of the reported compounds with the general formula $[M(L^1)(H_2O)_3]_2[M(H_2O)_6]\cdot(H_2O)_3$ (Table 2, entry 2; CCDC entry SIDJIZ was selected for comparison). This indicates that a phase transition or structure change of compound **1** to the corresponding SIDJIZ probably took place during the catalytic procedure. This phenomenon could not to be detected by TGA and IR measurements because of the similarities in the general formula. Experiments carried out with the model reaction and the recovered material showed that it can be reused at least four times with only a slight decrease in the catalytic activity (Table 3, entry 12). Because of the lack of porous channels within the structure of **1**, as well as the similar performance of the transformed recovered material, we envisage that the observed catalytic activity was revealed on the surface of the coordination polymer.

4. Conclusions

To summarize, in this work, we have continued our studies on the coordination chemistry of pseudopeptidic ligands. We synthesized a new three-dimensional Cu(II) coordination polymer with the tripodal trimesoyl-tris-glycine ligand. Compound **1** has a unique topological standard representation and can be considered as the first example of a 12-fold interpenetrated **ths** network. Synthetic-wise, a systematic study and comparison of all reported structures with this ligand and similar tripodal pseudopeptidic ligands showed that the choice of the Cu(II) starting material can have a large influence towards the self-assembly of the resulting product. Furthermore, **1** showed good catalytic activity towards the multicomponent synthesis of propargylamines under mild conditions; it was anticipated that the catalysis took place on the surface of the coordination polymer because of the lack of porosity. The recovered material could be reused for at least four cycles, however, PXRD studies pointed towards a structural change to a more favourable framework during the catalytic procedure. As such, the catalytic activity of **1** does not appear extremely promising for further efforts. Nevertheless, these initial results certainly demonstrate that coordination compounds with pseudopeptidic ligands could be tested as potential catalysts in organic reactions, provided that their structural framework remains stable. As a result of the above, our future efforts will thus focus on: (a) employing a variety of Cu(II) sources in more pseudopeptidic ligands to further study the self-assembly effect; (b) attempting to exploit the effect in order to get more interesting topologies; (c) identifying similar pseudopeptidic coordination compounds with higher stability in order to study their catalytic potential.

Supplementary Materials: The following are available online at www.mdpi.com/2073-4352/8/1/47/s1, Table S1: Crystal data and structure refinement for **1**; Figure S1: TGA graph for compound **1**, Figure S2: The IR spectrum of compound **1**, Figure S3: TGA graph for the recycled catalyst, Figure S4: TGA overlay of **1** (green) and the recycled catalyst (red), Figure S5: PXRD overlay of 1 pre- (red) and post- (black) catalysis. The simulated pattern of refcode SIDJIZ (mauve) is included for comparison; ^{1}H NMR of product propargylamines.

Acknowledgments: Vladislav Blatov and Davide Proserpio are acknowledged for helpful scientific discussions. Smaragda Lymperopoulou (University of Southampton, UK) is acknowledged for recording the PXRD data.

Author Contributions: Edward Loukopoulos and George E. Kostakis conceived and designed the experiments; Edward Loukopoulos and Alexandra Michail performed the experiments; Edward Loukopoulos, Alexandra Michail, and George E. Kostakis analysed the data; Edward Loukopoulos and George E. Kostakis wrote the paper.

Conflicts of Interest: The authors declare no conflict of interest.

References

1. Furukawa, H.; Cordova, K.E.; O'Keeffe, M.; Yaghi, O.M. The Chemistry and Applications of Metal-Organic Frameworks. *Science* **2013**, *341*, 1230444. [CrossRef] [PubMed]
2. Li, B.; Wen, H.M.; Cui, Y.; Zhou, W.; Qian, G.; Chen, B. Emerging Multifunctional Metal–Organic Framework Materials. *Adv. Mater.* **2016**, *28*, 8819–8860. [CrossRef] [PubMed]
3. Zhu, L.; Liu, X.Q.; Jiang, H.L.; Sun, L.B. Metal-Organic Frameworks for Heterogeneous Basic Catalysis. *Chem. Rev.* **2017**, *117*, 8129–8176. [CrossRef] [PubMed]
4. Yu, J.; Xie, L.-H.; Li, J.-R.; Ma, Y.; Seminario, J.M.; Balbuena, P.B. CO_2 Capture and Separations Using MOFs: Computational and Experimental Studies. *Chem. Rev.* **2017**, *117*, 9674–9754. [CrossRef] [PubMed]
5. Kirillov, A.M.; Karabach, Y.Y.; Kirillova, M.V.; Haukka, M.; Pombeiro, A.J.L. Topologically unique 2D heterometallic Cu II/Mg coordination polymer: Synthesis, structural features, and catalytic use in alkane hydrocarboxylation. *Cryst. Growth Des.* **2012**, *12*, 1069–1074. [CrossRef]
6. Liu, B. Metal–organic framework-based devices: Separation and sensors. *J. Mater. Chem.* **2012**, *22*, 10094. [CrossRef]
7. Wu, M.X.; Yang, Y.W. Metal–Organic Framework (MOF)-Based Drug/Cargo Delivery and Cancer Therapy. *Adv. Mater.* **2017**, *29*, 1606134. [CrossRef] [PubMed]
8. Cui, Y.; Chen, B.; Qian, G. Lanthanide metal-organic frameworks for luminescent sensing and light-emitting applications. *Coord. Chem. Rev.* **2014**, *273–274*, 76–86. [CrossRef]
9. Zhang, M.; Bosch, M.; Gentle, T., III; Zhou, H.-C. Rational design of metal–organic frameworks with anticipated porosities and functionalities. *CrystEngComm* **2014**, *16*, 4069. [CrossRef]
10. Burneo, I.; Stylianou, K.; Imaz, I.; Maspoch, D. The influence of the enantiomeric ratio of an organic ligand on the structure and chirality of Metal-Organic Frameworks. *Chem. Commun.* **2014**, *50*, 13829–13832. [CrossRef] [PubMed]
11. Loukopoulos, E.; Chilton, N.F.; Abdul-Sada, A.; Kostakis, G.E. Exploring the Coordination Capabilities of a Family of Flexible Benzotriazole-Based Ligands Using Cobalt(II) Sources. *Cryst. Growth Des.* **2017**, *17*, 2718–2729. [CrossRef]
12. Lu, W.; Wei, Z.; Gu, Z.-Y.; Liu, T.-F.; Park, J.; Park, J.; Tian, J.; Zhang, M.; Zhang, Q.; Gentle, T.; et al. Tuning the structure and function of metal-organic frameworks via linker design. *Chem. Soc. Rev.* **2014**, *43*, 5561–5593. [CrossRef] [PubMed]
13. Anderson, S.L.; Stylianou, K.C. Biologically derived metal organic frameworks. *Coord. Chem. Rev.* **2017**, *349*, 102–128. [CrossRef]
14. An, J.; Fiorella, R.P.; Geib, S.J.; Rosi, N.L. Synthesis, structure, assembly, and modulation of the CO_2 adsorption properties of a zinc-adeninate macrocycle. *J. Am. Chem. Soc.* **2009**, *131*, 8401–8403. [CrossRef] [PubMed]
15. Tan, Y.X.; He, Y.P.; Zhang, J. Serine-based homochiral nanoporous frameworks for selective CO_2 uptake. *Inorg. Chem.* **2011**, *50*, 11527–11531. [CrossRef] [PubMed]
16. Martí-Gastaldo, C.; Warren, J.E.; Stylianou, K.C.; Flack, N.L.O.; Rosseinsky, M.J. Enhanced stability in rigid peptide-based porous materials. *Angew. Chem. Int. Ed.* **2012**, *51*, 11044–11048. [CrossRef] [PubMed]
17. Wu, C.D.; Hu, A.; Zhang, L.; Lin, W. A homochiral porous metal-organic framework for highly enantioselective heterogeneous asymmetric catalysis. *J. Am. Chem. Soc.* **2005**, *127*, 8940–8941. [CrossRef] [PubMed]
18. Wu, X.; Zhang, H.-B.; Xu, Z.-X.; Zhang, J. Asymmetric induction in homochiral MOFs: From interweaving double helices to single helices. *Chem. Commun.* **2015**, *51*, 16331–16333. [CrossRef] [PubMed]
19. Wang, C.; Zheng, M.; Lin, W. Asymmetric catalysis with chiral porous metal-organic frameworks: Critical issues. *J. Phys. Chem. Lett.* **2011**, *2*, 1701–1709. [CrossRef]
20. An, J.; Shade, C.M.; Chengelis-Czegan, D.A.; Petoud, S.; Rosi, N.L. Zinc-adeninate metal-organic framework for aqueous encapsulation and sensitization of near-infrared and visible emitting lanthanide cations. *J. Am. Chem. Soc.* **2011**, *133*, 1220–1223. [CrossRef] [PubMed]
21. McKinlay, A.C.; Morris, R.E.; Horcajada, P.; Férey, G.; Gref, R.; Couvreur, P.; Serre, C. BioMOFs: Metal-organic frameworks for biological and medical applications. *Angew. Chem. Int. Ed.* **2010**, *49*, 6260–6266. [CrossRef] [PubMed]

22. Dokorou, V.N.; Milios, C.J.; Tsipis, A.C.; Haukka, M.; Weidler, P.G.; Powell, A.K.; Kostakis, G.E. Pseudopeptidic ligands: Exploring the self-assembly of isophthaloylbisglycine (H2IBG) and divalent metal ions. *Dalton Trans.* **2012**, *41*, 12501–12513. [CrossRef] [PubMed]
23. Dokorou, V.N.; Powell, A.K.; Kostakis, G.E. Two pseudopolymorphs derived from alkaline earth metals and the pseudopeptidic ligand trimesoyl-tris-glycine. *Polyhedron* **2013**, *52*, 538–544. [CrossRef]
24. Morrison, C.N.; Powell, A.K.; Kostakis, G.E. Influence of Metal Ion on Structural Motif in Coordination Polymers of the Pseudopeptidic Ligand Terephthaloyl-bis-beta-alaninate. *Cryst. Growth Des.* **2011**, *11*, 3653–3662. [CrossRef]
25. Lymperopoulou, S.; Dokorou, V.N.; Tsipis, A.C.; Weidler, P.G.; Plakatouras, J.C.; Powell, A.K.; Kostakis, G.E. Influence of the metal salt on the self-assembly of isophthaloylbis-β-alanine and Cu(II) ion. *Polyhedron* **2015**, *89*, 313–321. [CrossRef]
26. Kostakis, G.E.; Casella, L.; Boudalis, A.K.; Monzani, E.; Plakatouras, J.C. Structural variation from 1D chains to 3D networks: A systematic study of coordination number effect on the construction of coordination polymers using the terepthaloylbisglycinate ligand. *New J. Chem.* **2011**, *35*, 1060–1071. [CrossRef]
27. Kostakis, G.E.; Casella, L.; Hadjiliadis, N.; Monzani, E.; Kourkoumelis, N.; Plakatouras, J.C. Interpenetrated networks from a novel nanometer-sized pseudopeptidic ligand, bridging water, and transition metal ions with cds topology. *Chem. Commun.* **2005**, *987*, 3859–3861. [CrossRef] [PubMed]
28. Duan, J.; Zheng, B.; Bai, J.; Zhang, Q.; Zuo, C. Metal-dependent dimensionality in coordination polymers of a semi-rigid dicarboxylate ligand with additional amide groups: Syntheses, structures and luminescent properties. *Inorganica Chim. Acta* **2010**, *363*, 3172–3177. [CrossRef]
29. Wisser, B.; Chamayou, A.-C.; Miller, R.; Scherer, W.; Janiak, C. A chiral C3-symmetric hexanuclear triangular-prismatic copper(ii) cluster derived from a highly modular dipeptidic N,N'-terephthaloyl-bis(S-aminocarboxylato) ligand. *CrystEngComm* **2008**, *10*, 461–464. [CrossRef]
30. Sun, R.; Li, Y.Z.; Bai, J.; Pan, Y. Synthesis, structure, water-induced reversible crystal-to-amorphous transformation, and luminescence properties of novel cationic spacer-filled 3D transition metal supramolecular frameworks from N,N',N''-tris(carboxymethyl)-1,3,5-benzenetricarboxamide. *Cryst. Growth Des.* **2007**, *7*, 890–894. [CrossRef]
31. Sun, R.; Wang, S.; Xing, H.; Bai, J.; Li, Y.; Pan, Y.; You, X. Unprecedented 4264 topological 2-D rare-earth coordination polymers from a flexible tripodal acid with additional amide groups. *Inorg. Chem.* **2007**, *46*, 8451–8453. [CrossRef] [PubMed]
32. Loukopoulos, E.; Griffiths, K.; Akien, G.; Kourkoumelis, N.; Abdul-Sada, A.; Kostakis, G. Dinuclear Lanthanide (III) Coordination Polymers in a Domino Reaction. *Inorganics* **2015**, *3*, 448–466. [CrossRef]
33. Kallitsakis, M.; Loukopoulos, E.; Abdul-Sada, A.; Tizzard, G.J.; Coles, S.J.; Kostakis, G.E.; Lykakis, I.N. A Copper-Benzotriazole-Based Coordination Polymer Catalyzes the Efficient One-Pot Synthesis of (N'-Substituted)-hydrazo-4-aryl-1,4-dihydropyridines from Azines. *Adv. Synth. Catal.* **2017**, *359*, 138–145. [CrossRef]
34. Loukopoulos, E.; Kallitsakis, M.; Tsoureas, N.; Abdul-Sada, A.; Chilton, N.F.; Lykakis, I.N.; Kostakis, G.E. Cu(II) Coordination Polymers as Vehicles in the A^3 Coupling. *Inorg. Chem.* **2017**, *56*, 4898–4910. [CrossRef] [PubMed]
35. Kulkarni, C.; Meijer, E.W.; Palmans, A.R.A. Cooperativity Scale: A Structure–Mechanism Correlation in the Self-Assembly of Benzene-1,3,5-tricarboxamides. *Acc. Chem. Res.* **2017**, *50*, 1928–1936. [CrossRef] [PubMed]
36. Dolomanov, O.V.; Blake, A.J.; Champness, N.R.; Schröder, M. OLEX: New software for visualization and analysis of extended crystal structures. *J. Appl. Crystallogr.* **2003**, *36*, 1283–1284. [CrossRef]
37. Sheldrick, G.M. SHELXT—Integrated space-group and crystal-structure determination. *Acta Crystallogr. Sect. A Found. Adv.* **2015**, *71*, 3–8. [CrossRef] [PubMed]
38. Sheldrick, G.M. SHELXS97, Program for the Solution of Crystal Structures. *Acta Crystallogr. Sect. A* **2008**, *64*, 112–122. [CrossRef] [PubMed]
39. Sheldrick, G.M. A short history of SHELX. *Acta Crystallogr. Sect. A* **2008**, *64*, 112–122. [CrossRef] [PubMed]
40. Spek, A.L. Single-crystal structure validation with the program PLATON. *J. Appl. Crystallogr.* **2003**, *36*, 7–13. [CrossRef]
41. Farrugia, L.J. WinGX suite for small-molecule single-crystal crystallography. *J. Appl. Crystallogr.* **1999**, *32*, 837–838. [CrossRef]

42. Macrae, C.F.; Edgington, P.R.; McCabe, P.; Pidcock, E.; Shields, G.P.; Taylor, R.; Towler, M.; Van De Streek, J. Mercury: Visualization and analysis of crystal structures. *J. Appl. Crystallogr.* **2006**, *39*, 453–457. [CrossRef]

43. Stiefel, E.I.; Brown, G.F. On the Detailed Nature of the Six-Coordinate Polyhedra in Tris(bidentate ligand) Complexes. *Inorg. Chem.* **1972**, *11*, 434–436. [CrossRef]

44. Addison, A.W.; Rao, T.N.; Reedijk, J.; van Rijn, J.; Verschoor, G.C. Synthesis, structure, and spectroscopic properties of copper(II) compounds containing nitrogen-sulphur donor ligands; the crystal and molecular structure of aqua[1,7-bis(*N*-methylbenzimidazol-2′-yl)-2,6-dithiaheptane] copper(II) perchlorate. *J. Chem. Soc. Dalton. Trans.* **1984**, 1349. [CrossRef]

45. Blatov, V.A.; Shevchenko, A.P.; Proserpio, D.M. Applied Topological Analysis of Crystal Structures with the Program Package ToposPro. *Cryst. Growth Des.* **2014**, *14*, 3576–3586. [CrossRef]

46. Hao, Y.; Wu, B.; Li, S.; Jia, C.; Huang, X.; Yang, X.-J. Coordination polymers derived from a flexible bis(pyridylurea) ligand: Conformational change of the ligand and structural diversity of the complexes. *CrystEngComm* **2011**, *13*, 215–222. [CrossRef]

47. Hsu, Y.F.; Lin, C.H.; Chen, J.D.; Wang, J.C. A novel interpenetrating diamondoid network from self-assembly of *N,N′*-di(4-pyridyl)adipoamide and copper sulfate: An unusual 12-fold, [6 + 6] mode. *Cryst. Growth Des.* **2008**, *8*, 1094–1096. [CrossRef]

48. Blatov, V.A.; Carlucci, L.; Ciani, G.; Proserpio, D.M. Interpenetrating metal organic and inorganic 3D networks: A computer-aided systematic investigation. Part I. Analysis of the Cambridge structural database. *CrystEngComm* **2004**, *6*, 378. [CrossRef]

49. Allen, F.H. The Cambridge Structural Database: A quarter of a million crystal structures and rising. *Acta Crystallogr. Sect. B Struct. Sci.* **2002**, *58*, 380–388. [CrossRef]

50. Lu, Z.; Xing, H.; Sun, R.; Bai, J.; Zheng, B.; Li, Y. Water stable metal-organic framework evolutionally formed from a flexible multidentate ligand with acylamide groups for selective CO_2 adsorption. *Cryst. Growth Des.* **2012**, *12*, 1081–1084. [CrossRef]

51. Min, T.; Zheng, B.; Bai, J.; Sun, R.; Li, Y.; Zhang, Z. Topology diversity and reversible crystal-to-amorphous transformation properties of 3D cobalt coordination polymers from a series of 1D rodlike dipyridyl-containing building blocks and a flexible tripodal acid with additional amide groups. *CrystEngComm* **2010**, *12*, 70–72. [CrossRef]

52. Zuo, C.; Bai, J.; Sun, R.; Li, Y. Synthesis, structure, novel topology and reversible crystal-to-amorphous transformation of calcium coordination polymers from a flexible tripodal acid with additional amide groups. *Inorg. Chim. Acta* **2012**, *383*, 305–311. [CrossRef]

53. Karmakar, A.; Oliver, C.L.; Roy, S.; Öhrström, L. The synthesis, structure, topology and catalytic application of a novel cubane-based copper(ii) metal-organic framework derived from a flexible amido tripodal acid. *Dalton Trans.* **2015**, *44*, 10156–10165. [CrossRef] [PubMed]

54. Chen, Z.; Liu, X.; Wu, A.; Liang, Y.; Wang, X.; Liang, F. Synthesis, structure and properties of an octahedral dinuclear-based Cu 12 nanocage of trimesoyltri(L-alanine). *RSC Adv.* **2016**, *6*, 9911–9915. [CrossRef]

55. Chen, Z.; Zhang, C.; Liu, X.; Zhang, Z.; Liang, F. Synthesis, structure, and properties of a chiral zinc(II) metal-organic framework featuring linear trinuclear secondary building blocks. *Aust. J. Chem.* **2012**, *65*, 1662–1666. [CrossRef]

56. Chen, Z.; Liu, X.; Zhang, C.; Zhang, Z.; Liang, F. Structure, adsorption and magnetic properties of chiral metal–organic frameworks bearing linear trinuclear secondary building blocks. *Dalton Trans.* **2011**, *40*, 1911. [CrossRef] [PubMed]

57. Peshkov, V.A.; Pereshivko, O.P.; Van der Eycken, E.V. A walk around the A³-coupling. *Chem. Soc. Rev.* **2012**, *41*, 3790. [CrossRef] [PubMed]

58. Fan, W.; Yuan, W.; Ma, S. Unexpected E-stereoselective reductive A(3)-coupling reaction of terminal alkynes with aldehydes and amines. *Nat. Commun.* **2014**, *5*, 3884. [CrossRef] [PubMed]

59. Kaur, S.; Kumar, M.; Bhalla, V. Aggregates of perylene bisimide stabilized superparamagnetic Fe_3O_4 nanoparticles: An efficient catalyst for the preparation of propargylamines and quinolines via C–H activation. *Chem. Commun.* **2015**, *51*, 16327–16330. [CrossRef] [PubMed]

60. Paioti, P.H.S.; Abboud, K.A.; Aponick, A. Catalytic Enantioselective Synthesis of Amino Skipped Diynes. *J. Am. Chem. Soc.* **2016**, *138*, 2150–2153. [CrossRef] [PubMed]

61. De, D.; Pal, T.K.; Neogi, S.; Senthilkumar, S.; Das, D.; Gupta, S.S.; Bharadwaj, P.K. A Versatile CuII Metal-Organic Framework Exhibiting High Gas Storage Capacity with Selectivity for CO_2: Conversion of CO_2 to Cyclic Carbonate and Other Catalytic Abilities. *Chem. Eur. J.* **2016**, *22*, 3387–3396. [CrossRef] [PubMed]

62. Arcadi, A.; Cacchi, S.; Cascia, L.; Fabrizi, G.; Marinelli, F. Preparation of 2,5-disubstituted oxazoles from *N*-propargylamides. *Org. Lett.* **2001**, *3*, 2501–2504. [CrossRef] [PubMed]

63. Nilsson, B.M.; Hacksell, U. Base-Catalyzed cyclization of *N*-propargylamides to oxazoles. *J. Heterocycl. Chem.* **1989**, *26*, 269–275. [CrossRef]

64. Chauhan, D.P.; Varma, S.J.; Vijeta, A.; Banerjee, P.; Talukdar, P. A 1,3-amino group migration route to form acrylamidines. *Chem. Commun.* **2014**, *50*, 323–325. [CrossRef] [PubMed]

65. Yamamoto, Y.; Hayashi, H.; Saigoku, T.; Nishiyama, H. Domino coupling relay approach to polycyclic pyrrole-2-carboxylates. *J. Am. Chem. Soc.* **2005**, *127*, 10804–10805. [CrossRef] [PubMed]

66. Zhao, Y.; Zhou, X.; Okamura, T.-A.; Chen, M.; Lu, Y.; Sun, W.-Y.; Yu, J.-Q. Silver supramolecule catalyzed multicomponent reactions under mild conditions. *Dalton Trans.* **2012**, *41*, 5889–5896. [CrossRef] [PubMed]

67. Sun, W.J.; Xi, F.G.; Pan, W.L.; Gao, E.Q. MIL-101(Cr)-SO_3Ag: An efficient catalyst for solvent-free A^3 coupling reactions. *J. Mol. Catal. A Chem.* **2016**, *430*, 36–42. [CrossRef]

68. Jayaramulu, K.; Datta, K.K.R.; Suresh, M.V.; Kumari, G.; Datta, R.; Narayana, C.; Eswaramoorthy, M.; Maji, T.K. Honeycomb porous framework of zinc(II): Effective host for palladium nanoparticles for efficient three-component (A^3) coupling and selective gas storage. *Chempluschem* **2012**, *77*, 743–747. [CrossRef]

69. Lili, L.; Xin, Z.; Jinsen, G.; Chunming, X. Engineering metal–organic frameworks immobilize gold catalysts for highly efficient one-pot synthesis of propargylamines. *Green Chem.* **2012**, *14*, 1710. [CrossRef]

70. Prat, D.; Hayler, J.; Wells, A. A survey of solvent selection guides. *Green Chem.* **2014**, *16*, 4546–4551. [CrossRef]

crystals

MDPI

Article

Anthracene-Based Lanthanide Metal-Organic Frameworks: Synthesis, Structure, Photoluminescence, and Radioluminescence Properties

Stephan R. Mathis II [1], Saki T. Golafale [1], Kyril M. Solntsev [2] and Conrad W. Ingram [1,*]

[1] Center for Functional Nanoscale Materials, Department of Chemistry, Clark Atlanta University, Atlanta, GA 30314, USA; stepmath30@gmail.com (S.R.M.); saki.golafale@students.cau.edu (S.T.G.)

[2] School of Chemistry and Biochemistry, Georgia Institute of Technology, Atlanta, GA 30332, USA; solntsev@gatech.edu

* Correspondence: cingram@cau.edu

Received: 11 December 2017; Accepted: 17 January 2018; Published: 22 January 2018

Abstract: Four anthracene-based lanthanide metal-organic framework structures (MOFs) were synthesized from the combination of the lanthanide ions, Eu^{3+}, Tb^{3+}, Er^{3+}, and Tm^{3+}, with 9,10-anthracenedicarboxylic acid (H_2ADC) in dimethylformamide (DMF) under hydrothermal conditions. The 3-D networks crystalize in the triclinic system with P-1 space group with the following compositions: (i) $\{[[Ln_2(ADC)_3(DMF)_4 \cdot DMF]]_n$, Ln = Eu (**1**) and Tb (**2**)\} and (ii) $\{[[Ln_2(ADC)_3(DMF)_2(OH_2)_2 \cdot 2DMF \cdot H_2O]]_n$, Ln = Er (**3**) and Tm (**4**)\}. The metal centers exist in various coordination environments; nine coordinate in (i), while seven and eight coordinate in (ii). The deprotonated ligand, ADC, assumes multiple coordination modes, with its carboxylate functional groups severely twisted away from the plane of the anthracene moiety. The structures show ligand-based photoluminescence, which appears to be significantly quenched when compared with that of the parent H_2ADC solid powder. Structure **2** is the least quenched and showed an average photoluminescence lifetime from bi-exponential decay of 0.3 ns. On exposure to ionizing radiation, the structures show radioluminescence spectral features that are consistent with the isolation of the ligand units in its 3-D network. The spectral features vary among the 3-D networks and appear to suggest that the latter undergo significant changes in their molecular and/or electronic structure in the presence of the ionizing radiation.

Keywords: lanthanide coordination polymers; crystal structures; metal-organic framework fluorescence; radioluminescence; lanthanide metal-organic framework; lanthanide anthracene dicarboxylate coordination polymers

1. Introduction

Metal-organic framework (MOF) structures that display linker-based luminescence characteristics have been receiving an increasing amount of attention in recent years. Their 2-D and 3-D networks offer the opportunity to fine-tune the luminescence characteristics of the organic linker molecules by isolating them in well-defined environments [1–6]. Of interest to us is the potential of MOFs as radioluminescent (scintillating) materials for the detection of ionization radiation, including neutrons, protons, and gamma rays. Advancement in the science of detecting ionizing radiation is of great significance in radiography, biological safety, medical devices, biochemical analysis, particle physics, astrophysics, and nuclear materials identification and monitoring. Anthracene, with its highly luminescent chromophore, has the highest light output (rated at 100%) among organic scintillators and exhibits fast luminescent lifetimes [7]. The radioluminescence is based on energy transitions of excitable electrons in the π-bonds. This organic molecule can discriminate between subatomic particles and gamma rays, since the delayed fluorescence it produces is dependent on the nature of the exciting/ionizing particles. However, the natural arrangement of anthracene as a bulk solid is not conducive to high scintillation efficiency, which is hampered by the tendency of the molecule to dimerize on exposure to ionizing radiation [8]. It is possible that this drawback can be overcome by isolating the anthracene chromophore as a structural component of MOFs, thereby minimizing its tendency to undergo dimerization and reducing the associated non-radiative relaxation pathways.

Research on MOFs as radioluminescent materials has only been recent. Allendorf, Doty, and coworkers showed that MOFs that were assembled from Zn(II) ions and the deprotonated 4,4′-*trans*-stilbenedicarboxylic acid (SDC) are a new and effective class of scintillation materials [9–11]. We recently reported the synthesis and radioluminescence behavior of ultra large pores Ln-SDC MOFs (Ln = Tm and Er) [12]. In all cases, unique features were observed in their ligand-based photoluminescence and radioluminescence spectra and lifetimes, corresponding to differences in the electronic and crystalline structure of each material. The materials demonstrated increased fluorescence lifetimes that were indicative of higher fluorescence quantum efficiency from the rigidified stilbene linker within their structures. Allendorf, Doty, and coworkers further investigated the radioluminescence behavior of previously known MOF structures that contain other scintillating molecules as linkers, including 4,4′-biphenyldicarboxylic acid (Zn-based IRMOF-10), 2,6-naphthalenedicarboxylic acid (Zn-based IRMOF-8), an AlOH-based structure, 135-trimethylbenzene (DUT-6), 5,5-(naphthalene-2,6-diyl)diisophthalic acid (NOTT-103-Zn), and 5,5′-(anthracene-9,10-diyl)diisophthalinic acid (PCN-14) [2].

We are interested in exploring the behavior of anthracene when its molecular units are isolated in dense 3-D MOF structures that contain little or no voids or channels. We sought to investigate whether the inter-linker spacing within the structures would be sufficiently large to reduce inter-chromophore coupling, while close enough to impart rigidity and stability to the structure, thus decreasing non-radiative pathways while maintaining structural integrity upon exposure to ultraviolet or ionizing radiation. The organic moiety, 9,10-anthracenedicarboxylate (ADC), was therefore chosen as a linker, as this ligand is shorter and more rigid than 5,5′-(anthracene-9,10-diyl)diisophthalate of Zn-PCN-14, the only anthracene-based MOF whose scintillation behavior has been reported to date [2].

The choice of metal ion is also of consideration. A significant number of anthracene-based CP and MOF structures containing d-block and s-block metal ions have been reported. Among them are Cd [13–16], Zn [17–21], Ni [22], Co [23], Mn [24], Ag [25] and Mg [26], many of which demonstrate ligand-based photoluminescence. We chose to use the lanthanide metal ions. In comparison to d-block metal ions, the lanthanide ions demonstrate a propensity for high and variable coordination numbers and their flexible coordination environment are more conducive to forming more condensed multidimensional structures. Reports on anthracene-based lanthanide MOFs are sparse. A 3-D structure based on ADC in coordination with La(III) ions was reported by Wang et al. [13]. Like in the case of the transition metals, ligand-based photoluminescence was observed from this structure. Recently, Calahorro reported the synthesis and magnetic properties of Ln–ADC-based MOFs (Ln = Pr^{3+}, Nd^{3+}, Gd^{3+}, Tb^{3+}, Dy^{3+}, Er^{3+}, and Yb^{3+}) [27].

This manuscript reports the synthesis and photoluminescence and radioluminescence behavior of four Ln–ADC MOFs that were synthesized under hydrothermal conditions from the combination of Eu^{3+} (**1**), Tb^{3+} (**2**), Er^{3+} (**3**), and Tm^{3+} (**4**) with H_2ADC. Compounds **1**, **3**, and **4** are new. Compounds **3** and **4** are isostructural, as are Compounds **1** and **2**, and these compounds are among those recently reported [27].

2. Experimental Method

2.1. Synthesis

Synthesis of 1 ($C_{63}H_{59}N_5O_{17}Eu_2$): A mixture of $Eu(NO_3)_3 \cdot 6H_2O$ (0.043 g), $ADCH_2$ (0.05 mmol, 0.013 g), and DMF/H_2O (10 mL/10 mL) was sealed in a 20 mL scintillation vial and heated to 105 °C for 72 h in a convection oven. The vial was cooled to room temperature and the colorless crystals were filtered and repeatedly washed with fresh DMF and vacuum filtered. (Yield = 88% based on ADCH$_2$). Elemental Anal. (%) C, 51.80; H, 4.17; N, 4.92. Calcd. (%) C, 51.77; H, 4.04; N, 4.79. FTIR (KBr pellet, cm^{-1}): 3341 br, 1562 s, 1450 m, 1327 s, 1284 m, 1176 w, 1105 w, 1029 w, 839 s, 794 m, 736 m, 686 s, 600 w, 468 m.

Synthesis of 2 (C$_{63}$H$_{59}$N$_5$O$_{17}$Tb$_2$): The synthesis procedure was the same as **1**, except Tb(NO$_3$)$_3$·6H$_2$O was used as lanthanide metal salt (0.045 g). (Yield = 86% based on ADCH$_2$). Elemental Anal. (%) C, 51.22; H, 4.04; N, 4.80. Calcd. (%) C, 51.27; H, 3.99; N, 4.74. FTIR (KBr pellet, cm^{-1}): 3341 br, 1562 s, 1450 m, 1327 s, 1284 m, 1176 w, 1105 w, 1029 w, 839 s, 794 m, 736 m, 686 s.

Synthesis of 3 (C$_{60}$H$_{58}$N$_4$O$_{19}$Er$_2$): The synthesis procedure was the same as **1**, except Er(NO$_3$)$_3$·6H$_2$O (0.044 g) was used as lanthanide metal salt. (Yield = 92% based on ADC). Elemental Anal. (%) C, 49.00; H, 3.98; N, 3.75. Calcd. (%) C, 50.70; H, 3.95; N, 4.69. FTIR (KBr pellet, cm^{-1}): 3341 br, 1562 s, 1450 m, 1327 s, 1284 m, 1176 w, 1105 w, 1029 w, 839 s, 794 m, 736 m, 686 s, 600 w, 468 m.

Synthesis of 4 (C$_{60}$H$_{58}$N$_4$O$_{19}$Tm$_2$): The synthesis procedure was the same as **1** except Tm(NO$_3$)$_3$·6H$_2$O (0.045 g) was used as the lanthanide metal salt. (Yield = 86% based on ADC). Elemental Anal. (%) C, 41.74; H, 3.56; N, 3.19. Calcd. (%) C, 48.70; H, 3.93; N, 3.79. FTIR (KBr pellet, cm^{-1}): 3341 br, 1562 s, 1450 m, 1327 s, 1284 m, 1176 w, 1105 br, 1029 w, 839 s, 794 m, 736 m, 686 s, 600 w, 468 m.

2.2. Characterization

Single crystal X-ray analysis (SCXA) was conducted on a Bruker APEX-II CCD diffractometer. A suitable crystal was isolated from the sample and mounted onto the instrument using Paratone Oil. Measurements were made at ω scans of 1° per frame for 40 s using Mo Kα radiation (fine-focus sealed tube, 45 kV, 30 mA). The structures were solved with the Superflip structure solution program, using the Charge Flipping solution method [28] and using Olex2 as the graphical interface [29]. The models were refined with version 2013-4 of ShelXL using least squares minimization [30]. The total number of runs and images was based on the strategy calculation from the program APEX2 [31]. Cell parameters were retrieved and refined using SAINT software [32]. Data reduction was performed using the SAINT software, which corrects for Lorentz polarization. All non-hydrogen atoms were refined anisotropically. Hydrogen positions were calculated geometrically and refined using the riding model. Powdered X-ray diffraction (PXRD) patterns were recorded on a Panalytical Empyrean Series 2 X-ray diffractometer. The X-ray source was a Cu Kα (λ = 1.5418 Å) with anode at a voltage of 45 kV and current of 40 mA. Diffraction patterns were recorded between the 2θ angles of 4°–40° with a step size of 0.026°. Simulated PXRD patterns were obtained from SCXA data using Mercury 3.1 software from Cambridge Crystal Structure Database (CCDC). Infrared measurements were recorded on a Bruker Alpha-P FTIR spectrophotometer (intensive pattern: m—medium, s—strong, w—weak). The sample was introduced into the spectrophotometer using KBr as a zero-background powder and measurements were acquired between 350 and 4000 cm^{-1}. Thermogravimetric analysis was conducted on a TA Instrument Q50 thermal analyzer. Approximately 4 mg of sample was heated at a rate of 5 °C/min from ambient temperature to 900 °C under airflow. Elemental analysis was performed by Atlantic Microlab, Norcross, GA, USA.

2.3. Photoluminescence and Radioluminescence

Room temperature solid-state photoluminescence measurement was conducted on Photon Technology International fluorometer equipped with a 75 Watts Xenon Arc Lamp, excitation and emission monochromators, and a photomultiplier detector. A powdered sample of each structure was smeared between quartz slides, and excitation and emission spectra were recorded. Fluorescence lifetimes of the same samples were measured using an Edinburgh Instruments time-correlated single photon counting (TCSPC) system. In this measurement, an excitation pulse diode laser (LDH-P-C-375, 372 nm) was used as excitation light sources. The detection system consisted of a high-speed microchannel plate photomultiplier tube (MCP-PMT, Hamamatsu R3809U-50) and TCSPC electronics. The decays curves were fitted by the polyexponential functions after deconvolution with the instrument response function (IRF).

Radioluminescence measurements were conducted using an ion beam induced luminescence (IBIL) method in the Ion Beam Laboratory, Sandia Laboratories, New Mexico, using published procedures [11,12]. The experimental setup and conditions involved a 2.5 MeV proton beam, a current density 12,000 nA/cm^2 with the sample under 4.3×10^{-6} Torr vacuum pressure and ambient temperature. The beam was focused onto the sample with a spot size estimated to be 120 μm × 175 μm. Data was collected using a fiber optics coupled CCD spectrophotometer.

3. Results and Discussion

3.1. Structure Description

Structure description of 1 *and* 2: The PXRD patterns of **1** (Eu) and **2** (Tb), along with the simulated profile of **2**, show peaks in identical 2θ positions for both (Electronic Supplementary Information (ESI) Figure S1). The MOFs also show identical PXRD profiles as the simulated pattern of **1**, which confirms that each sample crystallizes as a pure phase. Crystal structure data for single SCXA are presented in Table 1. Both PXRD and SCXA analyses show that **1** and **2** are isostructural, so detailed analysis is given for **1** only. The crystal structure of **1** refines in a triclinic P-1 space group as a 3-D coordination polymeric network, with minimum voids and no measurable porosity.

Metal coordination: Structure **1** consists of two crystallographically identical Eu atoms per unit cell. There are five ADC units and two DMF molecules surrounding each Eu1 atom (Figure 1a). The Eu1 atom has nine coordinates, all with Eu1–O bonds. Seven Eu–O bonds are with the oxygen atoms from the five ADC ligands and two Eu1–O bonds are with oxygen atoms of two DMF molecules, generating an irregular EuO$_9$ coordination polyhedron. A list of select bond lengths and angles for **1** and **2** is presented in Table S1. The Eu–O bond lengths range from 2.363(1) Å for Eu–O3 to 2.572(2) Å for Eu–O1 and are similar to those observed in related coordination polymers [33,34]. The two Eu atoms are bridged by the oxygen atoms of two ADC ligands, with O1 atoms coordinating to Eu1 and creating an intradinuclear Eu⋯Eu distance of 4.036 Å.

Table 1. Crystal structure and refinement data for **1–4**.

Structure	1	2	3	4
Formula	$C_{63}H_{59}N_5O_{17}Eu_2$	$C_{63}H_{59}N_5O_{17}Tb_2$	$C_{60}H_{58}N_4O_{19}Er_2$	$C_{60}H_{58}N_4O_{19}Tm_2$
Density g cm^{-3}	1.645	1.681	1.697	1.703
μ/mm^{-1}	2.183	2.484	2.969	3.141
Formula Weight	1462.07	1475.99	1473.62	1476.96
Crystal System	triclinic	triclinic	triclinic	Triclinic
Space Group	P-1	P-1	P-1	P-1
a/Å	10.4144(5)	10.3706(9)	12.6026(14)	12.5949(3)
b/Å	11.4340(5)	11.3600(9)	14.9605(16)	14.9329(4)
c/Å	12.9875(6)	12.9850(11)	17.1024(19)	17.1174(5)
α/°	72.8203(11)	72.5410(10)	87.5905(14)	87.538(2)
β/°	89.9297(11)	89.7340(10)	69.7644(15)	69.7810(10)
γ/°	87.1572(12)	87.4110(10)	72.8494(14)	72.8160(10)
V/Å^3	1475.56(12)	1457.7(2)	2884.5(5)	2879.8(8)
Z	1	1	4	2
Z′	0.5	0.5	2	1
Θ_{min}/°	1.867	2.551	1.938	1.430
Θ_{max}/°	31.000	30.058	30.046	30.522
Measured Refl.	27,686	38,805	28,992	47,108
Independent Refl.	9326	8544	15,975	17,387
Refl. Used	8571	8036	11,489	13,514
R_{int}	0.0338	0.0338	0.0450	0.0527
Parameters	458	392	803	803
Restraints	95	24	98	65
Largest Peak	1.346	1.700	2.080	2.372
Deepest Hole	−0.999	−0.978	−1.169	−1.057
GooF	1.102	1.053	0.967	1.048
wR2(all data)	0.0753	0.0673	0.1202	0.0906
wR2	0.0680	0.660	0.1040	0.0814
R1(all data)	0.0322	0.0285	0.0750	0.0554
R1	0.0280	0.0264	0.0470	0.0374

Ligand coordination: The ADC unit coordinates in three modes, namely, $\mu_2:\eta_3$ (*bis*-bridging-chelating), $\mu_2:\eta_2$ (*bis*-bridging), and η_2 (*bis*-chelating). The bridging ligands link four metal atoms, whereas the *bis*-chelating ligands involve the linking of two. Various interconnected polymeric chains can be identified in the structure. Along the (100) direction, -(ADC-Eu$_2$)$_n$- chains are present, with the ADC units coordinating Eu atoms in *bis*-bidentate chelating mode, and with two adjacent Eu1 atoms bridged by oxygen atoms of the ligands carboxylate functional group (Figure 1b). Along the (010) direction, -(ADC-Eu$_2$)$_n$- chains are also present, with the ADC units coordinating Eu atoms in *bis*-bidentate bridging/chelating mode through the ligands' carboxylate groups. Each Eu atom is coordinated by three carboxylate oxygen atoms, and each carboxylate oxygen atom coordinates two Eu atoms. The -(ADC-Eu$_2$)$_n$- chains are also present along the *c* direction with the ADC units in *bis*-bidentate bridging mode; each carboxylate group coordinates two Eu atoms (Figure 1c). The nearest Eu$\cdots$Eu distance along the length of an ADC unit is 11.4 Å. Interestingly, the carboxylate groups are significantly twisted outside the plane of the anthracene moiety and measure 63.6° (C6-C2-C1-O2) for the *bis*-chelating, 85.8° (C11-C10-C9-O3) for the *bis*-bridging, and 68.7° for the *bis*-bridging/chelating ligands. The twisting of the carboxylate groups, as compared to a coplanar configuration, is consistent with reported DFT calculations on the parent H$_2$ADC molecule, which showed its potential energy to be at a minimum with a 60° rotation of the COOH group [35]. The interconnectivity of the chains through the Eu1 atoms creates a 3-D arrangement, with channels along the *b* axis and the coordinated DMF solvent inside. Though non-continuous solvent accessible voids constitute 11% of the structure, nitrogen adsorption analysis at 77 K on the activated sample of **1** shows no appreciable porosity.

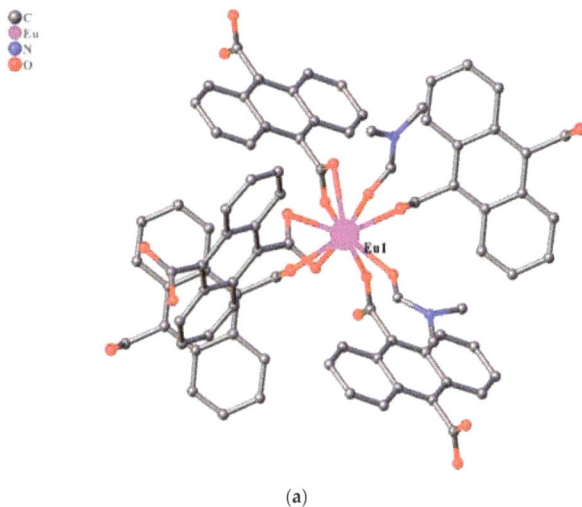

(a)

Figure 1. *Cont.*

(b)

(c)

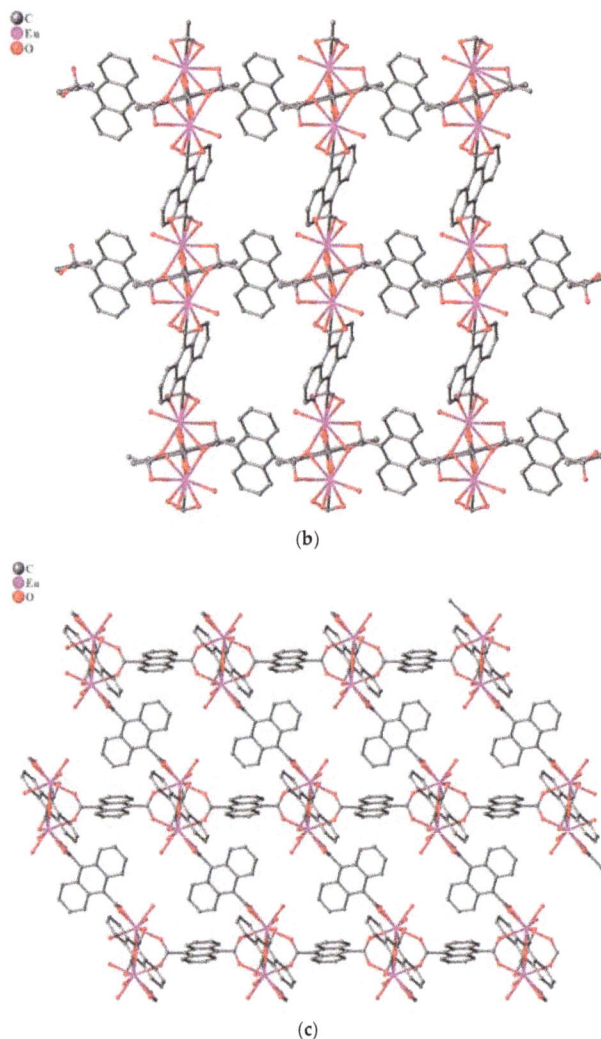

Figure 1. (a) Coordination environment of Eu1 atoms in 1. **(b)** Interconnectivity of *bis*-chelating chains along the (100) direction. **(c)** Interconnectivity of *bis*-bridging and *bis*-chelating chains along the (010) direction (Coordinated solvent and hydrogen atoms are omitted for clarity).

Structure description of 3 and 4: The PXRD patterns of **3 (Er)** and **4 (Tm)**, along with the simulated profile of **4**, show peaks in identical 2θ positions (ESI Figure S2). The structures show PXRD profiles that are identical to the simulated pattern of **4**, which confirms that they both crystallize as pure phases. Both PXRD and SCXA (Table 1) show that **3** and **4** are isostructural, so detailed analysis is given for **4** only. The structure is a 3-D network consisting of two crystallographically inequivalent Tm atoms, six ADC ligands, two coordinated, and two lattice, DMF molecules, and two coordinated, and one lattice, water molecules.

Metal Coordination: The coordination environment for each of the two crystallographically inequivalent Tm atoms is shown in Figure 2. The Tm1 atom has eight coordinates. Six are Tm1-O_{CARB} bonds with oxygen atoms from four ADC units, while the seventh and eight are Tm1-O_{H2O} and

Tm1-O$_{DMF}$ bonds, respectively (Figure 2a). The distorted Tm1O$_8$ coordination polyhedron has Tm1-O bond lengths ranging from 2.228(2) Å for Tm1-O10 to 2.512(3) Å for Tm1-O3, and O-Tm1-O bond angles ranging from 53.8(8)° to 89.1(4)°. The Tm2 atom coordinates seven oxygen atoms to form an irregular Tm2O$_7$ coordination polyhedron (Figure 2a). There are five Tm2-O$_{CARB}$ bonds from four ADC ligands, one Tm2-O$_{DMF}$ bond, and one Tm2-O$_{H2O}$. A list of selected bond lengths and angles are presented in Table S1. The Tm2-O bond lengths range between 2.219(3) Å for Tm2-O13 and 2.393(3) Å for Tm2-O5. The O-Tm2-O bond angles range from 55.7(3)° to 86.3(2)°. The Tm-O bond lengths are similar to those observed in related MOF structures [12]. The intradinuclear Tm1⋯Tm2 distance measures 4.687 Å.

Ligand coordination: Three coordination modes can be identified for the ADC linker, namely, bridging (μ_2:η_2), *bis*-chelating (η_2), and *bis*-monodentate (η_1). The *bis*-chelating ligands coordinate Tm1 and Tm2 atoms along both the *b* and *c* axes to create a 2-D "ladder-like" conformation, with the "ladder rungs" along the *b* axis (Figure 2b). The "ladder" structure is like that reported by Wang et al. in a related but different structure [13]. Along the *b* axis, the *bis*-chelating and *bis*-monodentate coordinating ADC units are coordinated to Tm1⋯Tm2 centers in an alternating arrangement. The *bis*-bridging ADC units, which are aligned along the *a* axis, intersect the 2-D ladder arrangement at the bimetallic Tm1⋯Tm2 centers to complete the 3-D network (Figure 2b). As in the case of **1** and **2**, the carboxylate groups are significantly twisted outside the plane of the anthracene moiety, with a twist angle of 62.2° (C30-C27-C26-O7) for the *bis*-bridging ligand and 77.6° (C3-C2-C1-O2) for the *bis*-chelating ligands. As discussed earlier, the twisting is dictated by the minimum energy conformation from the 60° rotation of the carboxylate, as shown by DFT calculations [34]. The uncoordinated oxygen atoms of the carboxylate groups on the *bis*-monodentate coordinating ADC units form strong hydrogen bonds with nearby coordinated water molecules, thus further stabilizing the ligand and providing restriction to its rotation (Figure 2c). The closest interchromophore distance is 12.5 Å. Narrow channels are present along the *b* direction. However, like **1**, nitrogen adsorption analysis at 77 K on an activated sample of **4** shows no appreciable porosity.

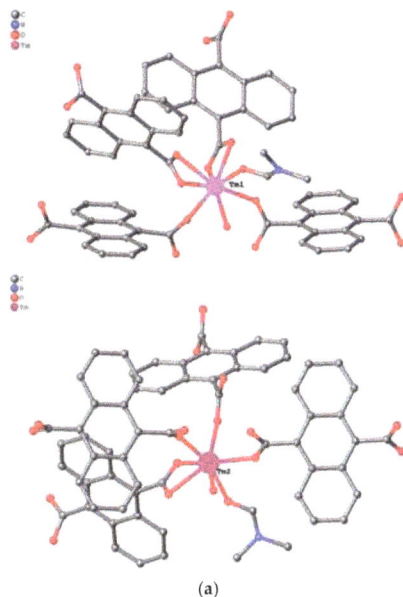

(a)

Figure 2. *Cont.*

(b)

(c)

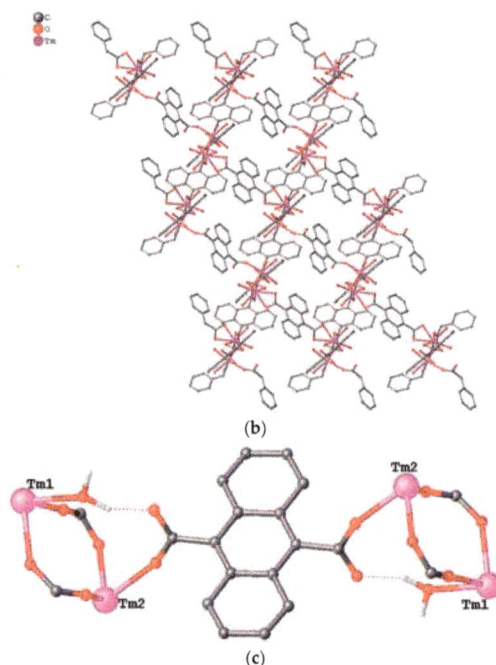

Figure 2. (**a**) The coordination environment of Tm1 and Tm2 in **4**. (**b**) Interconnectivity of two chains with the ADC units in *bis*-chelating and *bis*-monodentate coordination modes along the (010) direction. (**c**) Hydrogen bonding between ADC and coordinated water molecules. (Coordinated solvent molecules are omitted for clarity).

FTIR Analysis: The nature of the ADC units in the structures was further investigated by analysis of their Fourier transformed infrared spectra (FTIR) and a comparison with that of H_2ADC (ESI Figure S3). The band at 3448 cm^{-1} in all samples indicates the presence of O–H from adsorbed water on the MOFs and on H_2ADC. The 2925 cm^{-1} and 2967 cm^{-1} bands observed in H_2ADC were assigned to weak intramolecular O$\cdots$H bonds between non-planar C=O and H on the aromatic ring at the 1, 4, 5, and 8 carbon positions. These bands were not observed in **1–4**, and their absence is attributed to C=O coordination to the metal atom, thus limiting their interactions with aromatic H. The band observed at 1687 cm^{-1} in the spectrum of H_2ADC is attributed to the HO–C=O, with localized charges on the ligand's carboxylic acid functional groups. This band was not observed in **1–4**; instead, two individual bands were observed at 1601 cm^{-1} and 1551 cm^{-1}, which are attributed to variation in stretching vibrations of the C–O bonds in the three different ligand conformations. These observations indicate that the ligand is deprotonated (as ADC) within the MOF structures. The band at 1562 cm^{-1} in the MOF spectra is attributed to of metal-oxygen bonds [36].

Thermal Analysis: The thermal behavior of the structures was investigated by thermogravimetric analysis. The TGA curves of **1** and **2** show weight loss between 100 and 450 °C representing the loss of coordinated and uncoordinated DMF molecules (~30 wt %). Weight loss commencing around 440 °C (~40 wt %) is attributed to the loss of ADC units, and residue (~30 wt %) is attributed to lanthanide oxides. The TGA curves of **3** and **4** show small weight loss events up to 100 °C, attributed to the loss of H_2O molecules (~10 wt %). Weight loss up to 400 °C (~15 wt %) is attributed to loss of DMF. Weight loss event commencing around 400 °C (~45 wt %) is attributed to the loss of ADC units, and residue (~30 %) is attributed to lanthanide oxides.

3.2. Photoluminescence

The photoluminescence behavior of each compound was investigated. The room temperature solid-state photoluminescence emission spectra of the structures along with that of solid H_2ADC are presented in Figure 3. The spectrum of Na_2ADC in dilute aqueous solution was also recorded for comparison (ESI Figure S5). The spectrum of Na_2ADC in aqueous solution shows two defined vibronic peaks: one with λ_{max} at 425 nm and a shoulder at 450 nm. This is similar to that reported for pure anthracene [37], except that a smaller left shoulder peak expected at ~400 nm was not defined. The MOF structures (except **3**) show emission spectra with distinct vibronic peaks that are similar in profile to those observed for Na_2ADC in aqueous solution, thus suggesting that the emission is linker-based.

The emission peaks from the structures are within the 400–600 nm region, with their wavelength maxima (λ_{max}) observed between that of the Na_2ADC in dilute aqueous solution at 425 nm and that of the H_2ADC powder at 500 nm. The emission maxima of the structures are therefore red-shifted compared to the ADC sodium salt solution and blue-shifted compared to the H_2ADC powder. Further, a Stokes shift was observed among the structures as follows: 60 nm for **1** (380 nm $_{ex-max}$ to 440 nm $_{em-max}$), 65 nm for **2** (380 nm $_{ex-max}$ to 435 nm $_{em-max}$), 50 nm for **3** (389 nm $_{ex-max}$ to 430 nm $_{em-max}$) and 30 nm for **4** (400 nm $_{ex-max}$ to 430 nm $_{em-max}$), all of which are smaller than the 87 nm observed for H_2ADC. Except for **4**, the Stokes shift values are larger than the 41 nm observed in Zn-PCN-14, which contains the larger and rotatable anthracene liker, 5,5'-(anthracene-9,10-diyl)diisophthalinic acid (DPATC) [2]. With the exception of Structure **3**, the peaks are more defined than those observed in Zn-PCN-14. By comparison, the emission spectrum of H_2ADC is broad with less defined peaks and a larger Stokes shift (Figure 4). The broad spectral features of H_2ADC are likely ascribed to changes in the excited state geometry of the molecule due to the rotation of its –COOH groups to near coplanar conformation with the anthracene moiety. This can result in a lower energy excited state complex due to increased resonance, π overlap, and charge transfer interactions between the functional group and the ring system [34]. However, the ADC units in Structures **1–4** are deprotonated and are rigidified by coordination to the metal atoms. Rotation of the carboxylate groups is expected to be restricted as a result. This ridification, coupled with the separation of individual ADC units in the structures, will therefore reduce the level of interligand interactions and, by extension, reduce the extent of non-radiative relaxation pathways that would otherwise exist in solid forms of both anthracene and H_2ADC.

Within each structure, the cofacial alignment of ADC units in the (100) direction is interrupted by ADC units in the (010) direction. The closest cofacial distances are 14.5 Å in **1** and 12.5 Å in **4**, which are beyond the distance within which significant interchromophore coupling interactions among the phenyl rings of ADC units would be present [38]. The nearest distances between the planar face of one anthracene moiety and the hydrogen atoms on the edges of another, range between 3.689 (H12···C3) and 5.242 Å (H12···C7) for **1**, and between 3.397 (H4···C24) and 5.542 Å (H4···C21) for **4**. These edge-to-face distances are in the range within which C–H···π interactions are possible between the-orbitals of the hydrogens and the π system of the anthracene moieties. Such interactions could contribute to the non-radiative decay pathways and to the observed Stokes shifts. The broad featureless emission spectrum of **3** could be the result of more severe structural changes that brought ADC units closer on exposure to UV excitation. The possibility of the inductive effects of the Er atoms on the linker that can cause perturbation of the electronic transitions occurring in the ligand to result in more diffuse spectrum is also worthy of consideration and warrants further investigation.

Of note is that no luminescence spectral features from the lanthanide ions were detected in the visible region (for Eu^{3+} and Tb^{3+}) and were not measured in the near-infrared region (for Tm^{3+} and Er^{3+}) which is beyond the range of our standard fluorimeter. It is speculated, however, that following direct excitation the photoemissions of the latter two metals ions would be weak or non-existent due to their low molar absorption coefficient (typically lower that $10\ L\ mol^{-1}\ cm^{-1}$) [1–4].

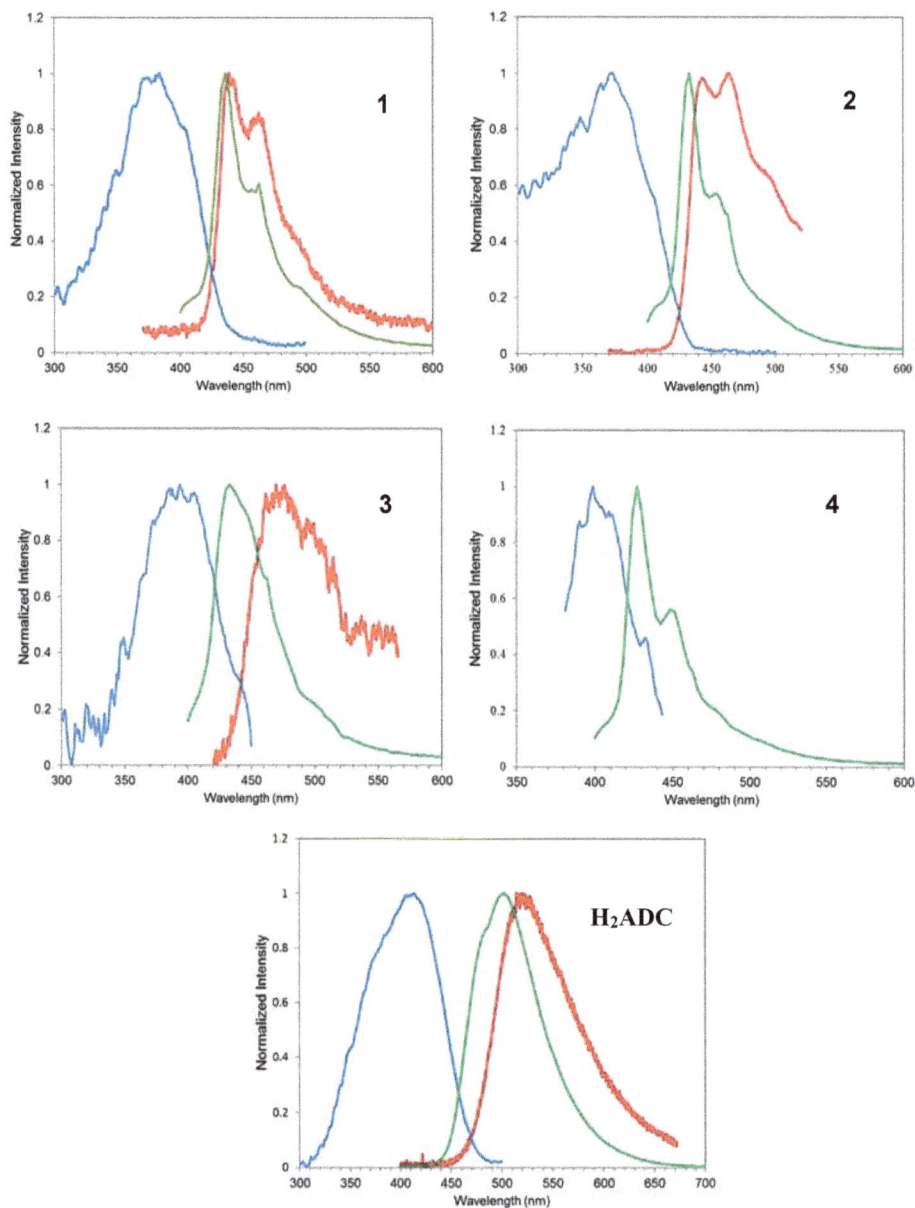

Figure 3. Photoluminescence excitation (**blue**), emission (**green**), and IBIL (**red**) spectra of **1–4** and H$_2$ADC. (Excitation wavelength was 380 nm for all emission spectra. Excitation spectra were monitored at emission wavelength of 435 nm for **1–4** and at 525 nm for H$_2$ADC. IBIL spectrum was not recorded for **4**).

Time-resolved photoluminescence decay: Time-resolved photoluminescence measurements were also acquired to further investigate the local environment of the anthracene units in the 3-D networks. Structures **1 Eu** (not shown), **3 (Er)**, and **4 (Tm)** yielded photoluminescence decay curves of low intensity that almost overlap with the instrument response function (IRF) (ESI Figure S6), while the decay curve for **2 (Tb)**, like that of H_2ADC, was quite distinct from the IRF (Figure 4). The decay curves were fitted with the biexponentional function, $I = \alpha_1 \exp\left(-\frac{\tau}{\tau_1}\right) + \alpha_2 \exp\left(-\frac{\tau}{\tau_2}\right)$, which corresponds to two different photo emissive rates, where I is the intensity, τ is the time, τ_1 and τ_2 are their corresponding excited state decay lifetimes, and α is the pre-exponential factor. The faster a radiative lifetime a major component has, the more τ_1 is attributed to emission from monomeric-like ADC units, and the more τ_2 is attributed to ADC units involved in coupling interactions, as observed for anthracene dimers. For Structure **2 (Tb)**, lifetimes $\tau_1 = 0.2$ ns and $\tau_2 = 0.5$ ns and weighted average lifetime $\tau_0 = 0.3$ ns are much shorter than the 4.9 ns, 16 ns, and 9.5 ns, respectively, we determined for H_2ADC (ESI Table S2). These lifetimes are also shorter than the average lifetime value of $\tau_0 = 2.0$ ns reported elsewhere for anthracene in monomeric isolated arrangements [39] and shorter than $\tau_0 = 5$ ns reported for Zn-PCN-14 [2].

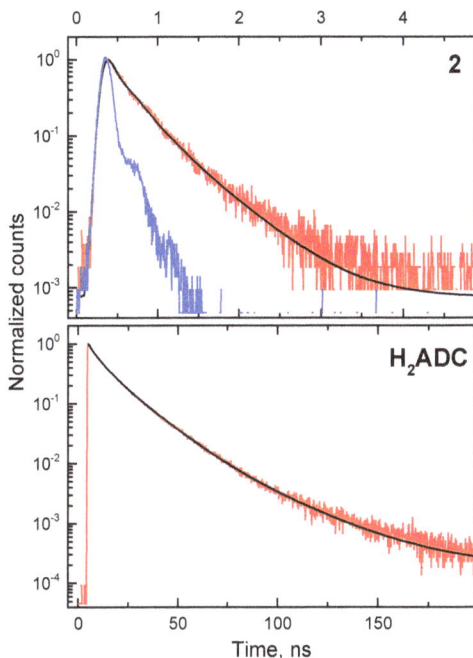

Figure 4. Photoluminescence decay curves of Structure **2** (top graph) and H_2ADC (blue = IRF, red = experimental data, black solid lines = model fit). (Excitation wavelength = 372 nm).

As discussed earlier, changes in the excited state geometry of bulk ligand molecules due to the rotation of its –COOH groups to near coplanar conformation with the anthracene moiety and reorganization towards end-to-face herringbone arrangement, which can facilitate excimer formation and strong interchromophore interactions, are quite likely. This could have contributed to the much longer lifetime (τ_2) compared to the MOF structures [8,34].

The short lifetimes observed for Structures **1–4** in comparison to those of H_2ADC and Zn-PCN-14 suggest that there is significant fluorescence quenching in the structures. Similar quenching of ligand fluorescence in complexes of lanthanides ions, including Tb^{3+} and Eu^{3+} have been previously observed [40] and is postulated to occur by the energy transfer between the ligand and the paramagnetic lanthanide ions via a cross-relaxation mechanism.

4. Radioluminescence

Radioluminescence (scintillation) is the emission of radiation after a material absorbs radiation with energy generally ≥ 10 eV that leads to π-electron ionization (Iπ) [7]. The ionization is followed by ion recombination (the recombining of secondary electrons with their parent electrons) that populates available singlet (S) and triplet (T) states. This is followed by non-radiative thermal deactivation to the lowest vibrational level of the first excited state S_1 before relaxation to lower electronic levels with an accompanying emission of radiation [41]. In this work, proton ion beam-induced luminescence (IBIL) spectroscopy was used to assess the radioluminescence spectral profile of the structures and H_2ADC, as this is known to simulate the production of recoil protons by elastic scattering of fast neutrons within an organic scintillator [2].

The IBIL emission spectral profiles for Structures **1–3** are compared with their respective photoluminescence spectra and to that of H_2ADC (Figure 3). The IBIL spectrum of **4** was not measured since the crystal faces were not of sufficiently large dimensions to fit the ion beam without penetration during data collection. The similarities in the IBIL spectral profile of **1** and **2** to their respective photoluminescence spectrum and to that of a dilute solution of Na_2ADC show that the IBIL is a product of the MOF crystal only and not of any H_2ADC impurities from synthesis or a result of any damage caused by the beam. Interestingly, for **1**, its IBIL spectrum shows two distinct vibronic peaks at 440 and 460 nm that almost overlap its photoluminescence spectrum (though a small red shift was observed). The overlap is consistent with the local environment of highly isolated ligand units in the structure remaining unaltered upon exposure to the proton beam [42]. For structures **2** (with $\lambda_{max\text{-}IBIL}$ at 445 nm) and **3** (with $\lambda_{max\text{-}IBIL}$ at 475 nm), there are significant red shifts in the IBIL spectrum of each when compared with their respective photoluminescence spectrum. This also translates into Stokes shift values (between the photoluminescence excitation λ_{max} and the $\lambda_{max\text{-}IBIL}$) of 65 (445–380 nm) for **2** and 85 nm (490–375 mm) for **3**, respectively. Interestingly also is that the IBIL emission peaks remained well defined in **2**, while they are broadened and featureless in **3**. These Stokes shift values are within close proximity of 78 nm reported for Zn-PCN-14 (which contains the DPATC linker) [2], but are much less than the 115 nm (515–400 nm) that we observed for H_2ADC (Figure 3). These Stokes shifts and peak broadening (in the case of **3**) suggest that the relaxation pathways described above may not be strictly observed due to changes in the molecular and/or electronic structures of materials on exposure to ionizing radiation. Such changes could possibly be a distortion of the chromophore environment, resulting in a shortening of inter-ligand distances and an increase in inter-chromophore interactions. It is notable also that, because of the significant Stokes shift, there was a relatively small spectral overlap between excitation and emission, and this is favorable for the use of the material as a scintillator in that self-absorption can be minimized [2].

5. Conclusions

The combination of lanthanide ions Eu^{3+}, Tb^{3+}, Er^{3+} and Tm^{3+} with the H_2ADC under hydrothermal conditions yielded four 3-D anthracene-based lanthanide MOFs in two different structure types; the Eu and Tb MOFs are isostructural, as are the Er and Tm MOFs. The deprotonated ligand, ADC, assumes multiple coordination modes in the structures and its carboxylate functional groups are severely twisted away from the plane of the anthracene moiety, which is consistent with its lowest energy conformation. The structures possess very narrow channels and show no appreciable porosity. The structures exhibit ligand-based photoluminescence that is significantly quenched. The Tb-containing MOF was least quenched and showed an average photoluminescence lifetime, τ_o, of 0.3 ns. On exposure to ionizing radiation, the structures also show ligand-based radioluminescence.

Supplementary Materials: The following are available online at http://www.mdpi.com/s1. Figure S1: Powder X-ray diffraction patterns of **1** and **2**, Figure S2: Powdered X-ray diffraction pattern of simulated **3** and **4**, Figure S3: FTIR spectra of **1–4** and H_2ADC, Figure S4: TGA curves of **1–4**, Figure S5: Photoluminescence emission spectrum of 0.1 M aqueous solution of Na_2ADC, Figure S6: Photoluminescence decay curves of structures **3** and **4**, Table S1: Selected bond lengths (Å) and angles (°) for **1–4**, Table S2: Photoluminescence lifetimes of **2–4** and H_2ADC.

Acknowledgments: This work was supported by the United States National Science Foundation, Grants No. HRD-0630456, HRD-1305041, the National Nuclear Security Administration, Grant No. NA0000979, and the Department of Energy, Grant No. DE-FE0022952. We thank John Bacsa and the Crystallography Laboratory at Emory University, Atlanta, GA, for conducting single crystal X-ray analyses of the structures. We also thank Elizabeth Auden and Khalid Hattar of Sandia National Laboratory Albuquerque, NM, for assistance in radioluminescence measurements. The authors are grateful to Rob Dickson and his group at Georgia Tech for the help with fluorescence lifetimes measurements. TGA curves, FTIR spectra, PXRD patterns, and the lists of bond lengths, bond angles, and other structural details are provided as ESI. CCDC No. 1046541, 1046540, 1046539, and 1046542 for **1**, **2**, **3**, and **4**, respectively, contains the supplementary crystallographic data for this paper. These data can be obtained free of charge via http://www.ccdc.cam.ac.uk/conts/retrieving.html (or from the CCDC, 12 Union Road, Cambridge CB2 1EZ, UK; Fax: +44-1223-336033; E-mail: deposit@ccdc.cam.ac.uk).

Author Contributions: All authors contributed equally to this work.

Conflicts of Interest: The authors declare no conflicts of interest.

References

1. Allendorf, M.D.; Bauer, C.A.; Bhakta, R.K.; Houk, R.J.T. Luminescent metal-organic frameworks. *Chem. Soc. Rev.* **2009**, *38*, 1330–1352. [CrossRef] [PubMed]
2. Perry, J.J., IV; Feng, P.L.; Meek, S.T.; Leong, K.; Doty, F.P.; Allendorf, M.D. Connecting structure with function in metal-organic frameworks to design novel photo- and radioluminescent materials. *J. Mater. Chem.* **2012**, *22*, 10235–10248. [CrossRef]
3. Kreno, L.E.; Leong, K.; Farha, O.K.; Allendorf, M.D.; Van Duyne, R.P.; Hupp, J.T. Metal-organic framework materials as chemical sensors. *Chem. Rev.* **2012**, *112*, 1105–1125. [CrossRef] [PubMed]
4. Rocha, J.; Carlos, L.D.; Almeida Paz, F.A.; Ananias, D. Luminescent multifunctional lanthanides-based metal–organic frameworks. *Chem. Soc. Rev.* **2011**, *40*, 926–940. [CrossRef] [PubMed]
5. Silva, C.G.; Corma, A.; Garciá, H. Metal-organic frameworks as semiconductors. *J. Mater. Chem.* **2010**, *20*, 3141–3156. [CrossRef]
6. Cui, Y.; Yue, Y.; Qian, G.; Chen, B. Luminescent functional metal-organic frameworks. *Chem. Rev.* **2012**, *112*, 1126–1162. [CrossRef] [PubMed]
7. Knoll, G.F. *Radiation Detection and Measurement*, 3rd ed.; John Wiley & Sons: Hoboken, NJ, USA, 2000; pp. 219–263.
8. Liu, J.Q. Crystal engineering of Cd(II) metal-organic frameworks bridged by dicarboxylates and N-donor coligands. *J. Coord. Chem.* **2011**, *64*, 1503–1512. [CrossRef]
9. Doty, F.; Bauer, C.; Skulan, A.; Grant, P.; Allendorf, M.D. Scintillating metal-organic frameworks: A new class of radiation detection materials. *Adv. Mater.* **2009**, *21*, 95–101. [CrossRef]
10. Bauer, C.A.; Timofeeva, T.V.; Settersten, T.B.; Patterson, B.D.; Liu, V.H.; Simmons, B.A.; Allendorf, M.D. Influence of connectivity and porosity on ligand-based luminescence in zinc metal-organic frameworks. *J. Am. Chem. Soc.* **2007**, *129*, 7136–7144. [CrossRef] [PubMed]
11. Feng, P.L.; Branson, J.V.; Hattar, K.; Vizkelethy, G.; Allendorf, M.D.; Doty, F.P. Designing metal-organic framework for radiation dectection. *Nucl. Inst. Methods Phys. Res. A* **2011**, *652*, 295–298. [CrossRef]
12. Mathis, S.R., II; Golafale, S.T.; Bacsa, J.; Steiner, A.; Ingram, C.W.; Doty, F.P.; Auden, E.; Hattar, K. Mesoporous stilbene-based lanthanide metal organic frameworks: Synthesis, photoluminescence and radioluminescence characteristics. *Dalton Trans.* **2017**, *46*, 491–500. [CrossRef] [PubMed]
13. Wang, J.; Hu, T.; Bu, X. Cadmium(II) and zinc(II) metal-organic frameworks with anthracene-based dicarboxylic ligands: Solvothermal synthesis, crystal structures, and luminescent properties. *CrystEngComm* **2011**, *13*, 5152–5161. [CrossRef]
14. Li, X.; Yang, L.; Zhao, L.; Wang, X.-L.; Shao, K.Z.; Su, Z.-M. Luminescent metal-organic frameworks with anthracene chromophores: Small-molecule sensing and highly selective sensing for nitro explosives. *Cryst. Growth Des.* **2016**, *16*, 4374–4382. [CrossRef]
15. Wang, J.; Liu, C.; Hu, T.; Chang, Z.; Li, C.; Yan, L.; Chen, P.; Bu, X.; Wu, Q.; Zhao, L.; et al. Zinc(II) coordination architectures with two bulky anthracene-based carboxylic ligands: Crystal structures and luminescent properties. *CrystEngComm* **2008**, *10*, 681–692. [CrossRef]
16. Liu, C.; Wang, J.; Chang, Z.; Yan, L.; Bu, X. Cadmium(II) coordination polymers based on a bulky anthracene-based dicarboxylate ligand: Crystal structures and luminescent properties. *CrystEngComm* **2010**, *12*, 1833–1841. [CrossRef]

17. Ma, S.; Wang, X.-S.; Collier, C.D.; Manis, E.S.; Zhou, H.-C. Ultramicroporous metal-organic framework based on 9,10-anthracenedicarboxylate for selective gas adsorption. *Inorg. Chem.* **2007**, *46*, 8499–8501. [CrossRef] [PubMed]

18. Chang, Z.; Zhang, A.-S.; Hiu, T.-L.; Bu, X.-H. Zn^{II} coordination poylmers based on 2,3,6,7-anthracenetetracarboxylic acid: Synthesis, structures, and luminescence properties. *Cryst. Growth Des.* **2009**, *9*, 4840–4846. [CrossRef]

19. Zhuang, J.; Friedel, J.; Terfort, A. The oriented and patterned growth of fluorescent metal-organic frameworks onto functionalized surfaces. *Belstein J. Nanotechnol.* **2012**, *3*, 570–578. [CrossRef] [PubMed]

20. Kojtari, A.; Carroll, P.; Ji, H. Metal organic framework (MOF) micro/nanopillars. *CrystEngComm* **2014**, *16*, 2885–2888. [CrossRef]

21. Gao, Q.; Xie, Y.-B.; Li, J.-R.; Yuan, D.-Q.; Yakovenko, A.A.; Sun, J.-H.; Zhou, H.-C. Tuning the formations of metal-organic frameworks by modification of ratio of reactant, acidity of reaction system, and use of a secondary ligand. *Cryst. Growth Des.* **2012**, *12*, 281–288. [CrossRef]

22. Ma, S.; Simmons, J.M.; Yuan, D.; Li, J.-R.; Weng, W.; Liu, D.-J.; Zhou, H.-C. A. Nanotubular metal-organic framework with permanent porosity: Structure analysis and gas sorption studies. *Chem. Commun.* **2009**, *27*, 4049–4051. [CrossRef] [PubMed]

23. Mu, Y.; Zhu, B.; Li, D.; Guo, D.; Zhao, J.; Ma, L. A new highly-connected 3D [Co$_4$(μ_3-OH)$_2$] cluster-based framework from different dicarboxylates and N-donor co-ligands: Synthesis, structure, and magnetic propert. *Inorg. Chem. Commun.* **2013**, *33*, 86–89. [CrossRef]

24. Liu, F.; Zhang, L.; Wang, R.; Sun, J.; Yang, J.; Chen, Z.; Wang, X.; Sun, D. Five MOFs with different topologies based on anthracene functionalized tetracarboxylic acid: Syntheses, structures, and properties. *CrysEngComm* **2014**, *16*, 2917–2928. [CrossRef]

25. Liu, C.; Chang, Z.; Wang, J.; Yan, L.; Bu, X.; Batten, S. A photoluminescent 3D silver(I) coordination polymer with mixed ligands anthracene-9,10-dicarboxylate and hexamethylenetetramine, showing binodal 4-connected (4^3·6^3)$_2$(4^2·6^2·8^2)$_3$ topology. *Inorg. Chem. Commun.* **2008**, *11*, 889–892. [CrossRef]

26. Chen, J.; Fan, Q.; Kitagawa, S. Synthesis, structures, and adsorption properties of two new magnesium coordination polymers. *Solid State Sci.* **2013**, *16*, 29–33. [CrossRef]

27. Calahorro, A.J.; Fernández, B.; Oyarzabal, I.; Seco, J.M.; Tian, T.; Fairen-Jimenez, D.; Colacio, E.; Rodríguez-Diéguez, A. Rare earth anthracenedicarboxylate metal-organic frameworks: Slow relaxation of magnetization of Nd^{3+}, Gd^{3+}, Dy^{3+}, Er^{3+} and Yb^{3+} based materials. *Dalton Trans.* **2016**, *45*, 591–598. [CrossRef] [PubMed]

28. Palatinus, L.; Chapuis, G. *SUPERFLIP*—A computer program for the solution of crystal structures by charge flipping in arbitrary dimensions. *J. Appl. Cryst.* **2007**, *40*, 786–790. [CrossRef]

29. Dolomanov, O.V.; Bourhis, L.J.; Gildea, R.J.; Howard, J.A.K.; Puschmann, H. *OLEX2*—A complete structure solution, refinement and analysis program. *J. Appl. Cryst.* **2009**, *42*, 339–341. [CrossRef]

30. Sheldrick, G.M. A short history of SHELX. *Acta Cryst.* **2008**, *A64*, 339–341.

31. *APEX2*; Bruker AXS Inc.: Madison, WI, USA, 2012.

32. *SAINT*; Bruker AXS Inc.: Madison, WI, USA, 2013.

33. Wang, J.; Wang, J.; Li, Y.; Jiang, M.; Zhang, L.; Wu, P. A europium(III)-based metal–organic framework as a naked-eye and fast response luminescence sensor for acetone and ferric iron. *New J. Chem.* **2016**, *40*, 8600–8606. [CrossRef]

34. Wen, G.-X.; Wu, Y.-P.; Dong, W.-W.; Zhao, J.; Li, D.-S.; Zhang, J. An Ultrastable europium(III)-organic framework with the capacity of discriminating Fe^{2+}/Fe^{3+} ions in various solutions. *Inorg. Chem.* **2016**, *55*, 10114–10117. [CrossRef] [PubMed]

35. Werner, T.C.; Hercules, D.M. Fluorescence of 9-anthroic acid and its esters. Environmental effects on excited-state behavior. *J. Phys. Chem.* **1969**, *73*, 2005–2011. [CrossRef]

36. Dametto, P.; Siqueira, A.; Carvalho, C.; Ionashiro, M. Synthesis, characterization and thermal studies on solid state 3-methoxybenzoate of lighter trivalent lanthanides. *Eclética Química* **2007**, *32*, 17–21. [CrossRef]

37. Clarke, H.B.; Northrop, D.C.; Simpson, O. The Scintillation Phenomenon in Anthracene I. Radiation Damage. *Prog. Phys. Soc.* **1962**, *79*, 366–372. [CrossRef]

38. Nag, A.; Kutty, T. The light induced valence change of europium in Sr$_2$SiO$_4$: Eu involving transient crystal structure. *J. Mater. Chem.* **2004**, *14*, 1598–1604. [CrossRef]

39. Hinoue, T.; Shigenoi, Y.; Sugino, M.; Mizobe, Y.; Hisaki, I.; Miyata, M.; Tohnai, N. Regulation of π-stacked anthracene arrangement for fluorescence modulation of organic solid from monomer to excited oligomer emission. *Chem. Eur. J.* **2012**, *18*, 4634–4643. [CrossRef] [PubMed]
40. Birks, J.B. *The Theory and Practice of Scintillation Counting*, 1st ed.; Pargamon: Oxford, UK, 1964.
41. King, T.A.; Voltz, R. The time dependence of scintillation intensity in aromatic materials. *Proc. R. Soc. Lond. Ser. A* **1966**, *269*, 424–439. [CrossRef]
42. Birks, J.B. Energy Transfer in organic phosphorsm. *Phys. Rev.* **1954**, *94*, 1567. [CrossRef]

crystals

MDPI

Review

Multifunctional Aromatic Carboxylic Acids as Versatile Building Blocks for Hydrothermal Design of Coordination Polymers

Jinzhong Gu [1,*], Min Wen [1], Xiaoxiao Liang [1], Zifa Shi [1], Marina V. Kirillova [2] and Alexander M. Kirillov [2,*]

[1] Key Laboratory of Nonferrous Metal Chemistry and Resources Utilization of Gansu Province, State Key Laboratory of Applied Organic Chemistry and College of Chemistry and Chemical Engineering, Lanzhou University, Lanzhou 730000, China; wenm17@lzu.edu.cn (M.W.); liangxx15@lzu.edu.cn (X.L.); shizf@lzu.edu.cn (Z.S.)

[2] Centro de Química Estrutural, Instituto Superior Técnico, Universidade de Lisboa, Av. Rovisco Pais, 1049-001 Lisbon, Portugal; kirillova@tecnico.ulisboa.pt

* Correspondence: gujzh@lzu.edu.cn (J.G.); kirillov@tecnico.ulisboa.pt (A.M.K.)

Received: 15 January 2018; Accepted: 29 January 2018; Published: 3 February 2018

Abstract: Selected recent examples of coordination polymers (CPs) or metal-organic frameworks (MOFs) constructed from different multifunctional carboxylic acids with phenyl-pyridine or biphenyl cores have been discussed. Despite being still little explored in crystal engineering research, such types of semi-rigid, thermally stable, multifunctional and versatile carboxylic acid building blocks have become very promising toward the hydrothermal synthesis of metal-organic architectures possessing distinct structural features, topologies, and functional properties. Thus, the main aim of this mini-review has been to motivate further research toward the synthesis and application of coordination polymers assembled from polycarboxylic acids with phenyl-pyridine or biphenyl cores. The importance of different reaction parameters and hydrothermal conditions on the generation and structural types of CPs or MOFs has also been highlighted. The influence of the type of main di- or tricarboxylate ligand, nature of metal node, stoichiometry and molar ratio of reagents, temperature, and presence of auxiliary ligands or templates has been showcased. Selected examples of highly porous or luminescent CPs, compounds with unusual magnetic properties, and frameworks for selective sensing applications have been described.

Keywords: coordination polymers; metal-organic frameworks; crystal engineering; hydrothermal synthesis; carboxylic acids

1. Introduction and Scope

In recent years, various crystalline metal-organic architectures (MOAs) including coordination polymers (CPs) or metal-organic frameworks (MOFs) have been an object of very intense research that spans from the fields of crystal design and engineering to chemistry of functional materials [1–15]. In particular, a very interesting research direction concerns the search for new and versatile organic building blocks that can be applied for the design of unusual metal-organic architectures with desirable structural features and notable functional properties [16–19]. Despite considerable progress achieved in this field, the assembly of coordination polymers or metal-organic frameworks in a predictable way is often a difficult task. This is mainly because the assembly of such compounds can depend on various factors, such as the nature and coordination properties of metal nodes [20,21], connectivity and type of organic building blocks [22–24], reaction conditions and stoichiometry [25,26], and effects of templates [27–29] or supporting ligands [30,31].

A high diversity of aromatic polycarboxylic acids has been extensively applied as multifunctional building blocks in designing novel metal-organic networks [32,33]. Among such building blocks, flexible ligands containing biphenyl and phenyl-pyridine cores with a varying number and position of carboxylic groups as well as distinct locations of *N*-pyridyl functionality have attracted a special interest [34,35]. It can be justified by a possibility of two adjacent phenyl and/or pyridine rings to rotate around the C–C single bond and thus conform to a coordination environment of metal nodes. Besides, the presence of several carboxylic groups with a varying degree of deprotonation in addition to an optional *N*-pyridyl functionality can provide multiple and distinct coordination sites, thus leading to different coordination fashions and resulting in the assembly of structurally distinct coordination polymers [36–38]. Furthermore, depending on a deprotonation degree and crystal packing arrangement, these aromatic polycarboxylate ligands can behave as good H-bond acceptors and donors, thus furnishing an extra stabilization of metal-organic structures and facilitating their crystallization.

Hence, the main objective of the present work consists in highlighting selected recent examples of coordination polymers that were hydrothermally assembled from a series of multifunctional carboxylic acids with phenyl-pyridine or biphenyl cores (Scheme 1). These carboxylic acids are still very poorly explored toward the design of CPs or MOFs, but can constitute an interesting type of semi-rigid, thermally stable, multifunctional, and versatile building blocks in crystal engineering research. Thus, the present study briefly discusses the general aspects of hydrothermal synthesis of selected coordination polymers derived from the aromatic carboxylic acids shown in Scheme 1. Some of them represent isomeric biphenyl tricarboxylate blocks (H_3bptc and H_3btc), while other are isomeric phenyl-pyridine tricarboxylate blocks (H_3cptc, H_3dcppa, and H_3cpta). The study also highlights the influence of various parameters (main ligand type, metal node, molar ratio and stoichiometry, temperature, presence of auxiliary ligand or template) on structural diversity of the obtained products. For selected examples of CPs, functional properties and applications are also highlighted.

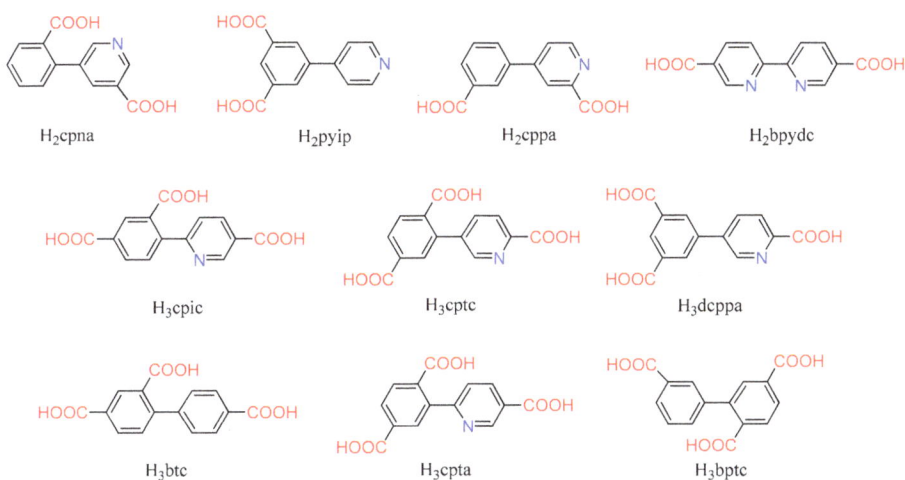

Scheme 1. Ten selected multifunctional carboxylic acids used as building blocks for the design of CPs or MOFs.

2. Hydrothermal Synthesis and Structural Diversity of Coordination Polymers

2.1. Advantages of Hydrothermal Synthesis

Hydrothermal synthesis is commonly applied toward the design of metal-organic networks [39–41] and refers to the synthesis and crystallization of coordination compounds that occur under hydrothermal conditions, typically in a hermetically sealed aqueous solution at elevated temperatures and pressures. The hydrothermal synthesis features a number of important advantages over other common methods for preparing CPs, namely: (i) high reactivity of reactants and unique synthetic conditions in terms of a combination of pressures and temperatures; (ii) growth of good quality single crystals (Figure 1) or microcrystalline phases with no need for additional work-up and purification; (iii) possible control of solution or interface reactions, formation of metastable and unique structures that cannot be generated by other methods; (iv) use of water as a green organic-solvent-free reaction medium that can also aid crystallization by supplying labile H_2O ligands to complete coordination environment of metal nodes; and (v) relative simplicity of the equipment.

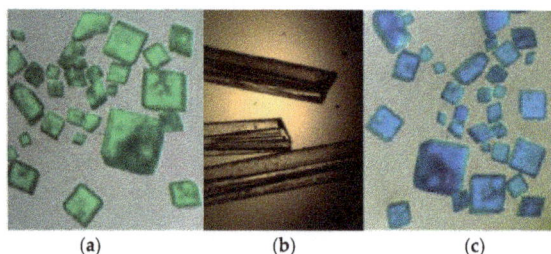

| (a) | (b) | (c) |

Figure 1. Images showing examples of single crystals of Ni (**a**), Cd (**b**), and Cu (**c**) coordination polymers generated hydrothermally.

For coordination polymers driven by multifunctional carboxylic acids with phenyl-pyridine or biphenyl cores (Scheme 1), typical synthetic procedure begins with mixing, in water at ambient temperature and under constant stirring, a metal nitrate or chloride salt, a main carboxylic acid building block, and an auxiliary ligand (optional) [42–44]. The obtained mixture is then treated with sodium hydroxide as a typical base to adjust the solution pH value in the range of 5–7. Then, the reaction mixture is sealed in a Teflon-lined stainless steel autoclave and subjected to the hydrothermal treatment at 80–210 °C for 2 or 3 days in an oven, followed by gradual cooling to ambient temperature at a rate of 10 °C/h (Figure 2). The autoclaves are opened after being kept at ambient temperature for 24 h. The obtained crystalline solids are filtered off and washed (optional) or isolated manually to furnish a coordination polymer product (Figure 1).

| (a) | (b) |

Figure 2. Images of the Teflon-lined stainless steel autoclaves (**a**) and an oven with temperature control (**b**) typically applied for the hydrothermal generation of CPs.

2.2. Effect of Building Block Type

The type of the main carboxylic acid ligand (Scheme 1) is one of the structure-defining factors during the hydrothermal synthesis of CPs. Selected examples of different metal-organic networks that were obtained under similar reaction conditions are summarized in Table 1. For example, the use of different dicarboxylic acids (H_2cpna, H_2cppa, or H_2bpydc) as main building blocks and 1,10-phenanthroline as an auxiliary ligand led to the generation of distinct manganese(II) derivatives **1**−**3** (Figure 3), the structures of which range from a 1D ladder [Mn(μ_3-cpna)(phen)(H_2O)]$_n$ (**1**) and 1D zigzag chain [Mn(μ-cppa)(phen)(H_2O)]$_n$ (**2**) to a 3D MOF [Mn(μ_4-bpydc)(phen)]$_n$ (**3**). The use of a cobalt(II) metal source in combination with the isomeric H_2cpna or H_2cppa ligands and 2,2′-bipyridyl resulted in the assembly of a 2D metal-organic layer [Co(μ_3-cpna)(2,2′-bpy)(H_2O)]$_n$ (**4**) or a 1D zigzag chain {[Co(μ-cppa)(2,2′-bpy)(H_2O)]·H_2O}$_n$ (**5**). Similar structure-defining influence of tricarboxylic acid building blocks can be observed in other zinc(II) (**6**, **7**) and manganese(II) (**8**, **9**) coordination compounds (Table 1).

Table 1. Selected examples of coordination polymers (CPs) showing an effect of main carboxylate ligand on product structure.

Compound	Formula	Ligand	Structure	Reference
1	[Mn(μ_3-cpna)(phen)(H_2O)]$_n$	H_2cpna	1D ladder chain	[42]
2	[Mn(μ-cppa)(phen)(H_2O)]$_n$	H_2cppa	1D zigzag chain	[44]
3	[Mn(μ_4-bpydc)(phen)]$_n$	H_2bpydc	3D MOF	[45]
4	[Co(μ_3-cpna)(2,2′-bpy)(H_2O)]$_n$	H_2cpna	2D layer	[42]
5	{[Co(μ-cppa)(2,2′-bpy)(H_2O)]·H_2O}$_n$	H_2cppa	1D zigzag chain	[44]
6	[Zn$_3$(μ_3-cptc)$_2$(H_2O)$_6$]$_n$	H_3cptc	1D ladder chain	[46]
7	{[Zn$_3$(μ_5-dcppa)$_2$(H_2O)$_4$]·2H_2O}$_n$	H_3dcppa	3D MOF	[47]
8	[Mn(μ-Hdcppa)(phen)(H_2O)]$_2$·2H_2O	H_3dcppa	0D dimer	[47]
9	{[Mn(μ_4-Hcpta)(phen)]·4H_2O}$_n$	H_3cpta	3D MOF	[48]

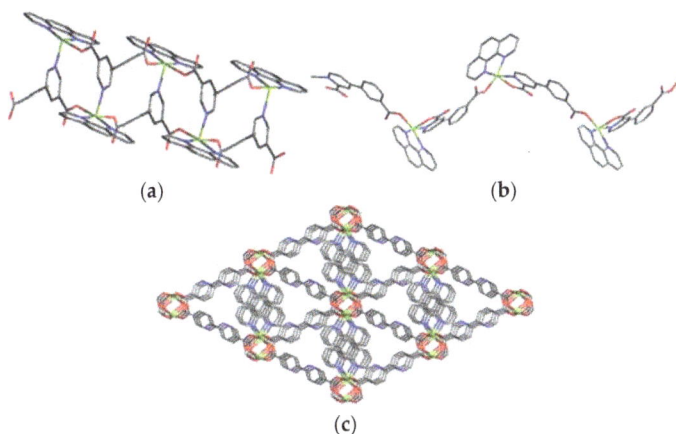

(a)

(b)

(c)

Figure 3. (a) 1D ladder chain in **1**. (b) 1D zigzag chain in **2**. (c) 3D metal-organic framework in **3**. Adapted from [42,44,45].

2.3. Effect of Metal Source

The type of metal node also plays an important structure-defining role in the hydrothermal generation of coordination polymers. This is primarily associated with different coordination behavior and ligand affinity of distinct metal centers, their charges and ionic radii. Selected examples of CPs assembled under identical reaction conditions but using different metal sources are collected in Table 2. In particular, an interesting series of compounds **14−16** can be built from H_3btc and phen ligands by using different metal(II) chlorides, namely a 1D chain{[Cd(μ_3-Hbtc)(phen)(H_2O)]·H_2O}$_n$ (**14**), a 3D MOF [Pb$_3$(μ_4-Hbtc)$_2$(phen)]$_n$ (**15**), and a 0D monomer [Ni(Hbtc)$_2$(phen)$_2$(H_2O)]·2H_2O (**16**). Notably, despite the diversity of these structures, they all feature a monoprotonated tricarboxylic acid block, Hbtc$^{2−}$.

Apart from the nature of metal, the type of anion in a starting metal salt can also influence the resulting structure. For example, samarium(III) coordination polymers {[Sm(Hcpna)(μ_4-cpna)(phen)]$_2$·H_2O}$_n$ (3D net, **17**) and {[Sm(Hcpna)(μ_4-cpna)(phen)]$_2$·2H_2O}$_n$ (1D chain, **18**) were obtained under exactly the same conditions but using Sm(III) nitrate or chloride, respectively. MOF **17** reveals a very intricate structure, wherein the Sm$_2$ dimeric units are linked by the μ_4-cpna$^{2−}$ ligands forming a dodecanuclear Sm$_{12}$ macrocycle (Figure 4a) that adopts a chair conformation. These Sm$_{12}$ units are then connected with six adjacent rings by corner-forming 2D layer motifs (Figure 4b), which are further linked by the coordination interaction with the cpna$^{2−}$ blocks to furnish a very complex 3D framework (Figure 4c).

Table 2. Selected examples of CPs showing an effect of metal source on product structure.

Compound	Formula	Metal Source	Structure	Reference
10	[Co(μ-cppa)(phen)(H_2O)]$_n$	CoCl$_2$·6H_2O	1D zigzag chain	[44]
11	{[Cd$_3$(μ_3-cppa)$_3$(phen)$_2$]·4H_2O}$_n$	CdCl$_2$·H_2O	3D MOF	[44]
12	{[Y$_2$(μ_4-cpna)$_3$(phen)$_2$(H_2O)]·H_2O}$_n$	Y(NO$_3$)$_3$·6H_2O	3D MOF	[43]
13	[Tm(μ_3-cpna)(phen)(NO$_3$)]$_n$	Tm(NO$_3$)$_3$·6H_2O	1D double chain	[43]
14	{[Cd(μ_3-Hbtc)(phen)(H_2O)]·H_2O}$_n$	CdCl$_2$·H_2O	1D chain	[49]
15	[Pb$_3$(μ_4-Hbtc)$_2$(phen)]$_n$	PbCl$_2$	3D MOF	[49]
16	[Ni(Hbtc)$_2$(phen)$_2$(H_2O)]·2H_2O	NiCl$_2$·6H_2O	0D monomer	[50]
17	{[Sm(Hcpna)(μ_4-cpna)(phen)]$_2$·H_2O}$_n$	Sm(NO$_3$)$_3$·6H_2O	3D MOF	[43]
18	{[Sm(Hcpna)(μ_4-cpna)(phen)]$_2$·2H_2O}$_n$	SmCl$_3$·6H_2O	1D chain	[43]

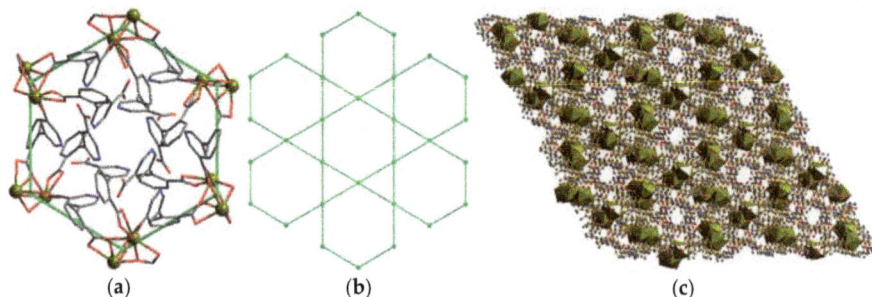

(a)   (b)   (c)

Figure 4. Structural fragments of MOF **17**. (**a**) Hexagonal Sm$_{12}$ macrocycle; green balls are Sm$_2$ units. (**b**) Interconnection of hexagonal macrocycles into a 2D layer motif; green balls are Sm$_2$ units. (**c**) 3D metal-organic framework. Adapted from [43].

2.4. Effect of Reagents Molar Ratio

In the synthesis of CPs, a proportion between metal node and main carboxylate ligand can be easily modified, what can cause a change of the coordination number of metal ions and affect the resulting structure. In addition, change of the molar ratio between main building block and alkali metal hydroxide used as a pH-regulator can result in a partial or full deprotonation of polycarboxylic acid ligand. As shown in Table 3, both 3D MOFs {[Co$_3$(μ_4-btc)$_2$(μ-H_2O)$_2$(py)$_4$(H_2O)$_2$]·(py)$_2$}$_n$ (**19**) and {[Co$_{3.5}$(μ_6-btc)$_2$(μ_3-OH)(py)$_2$(H_2O)$_3$]·H_2O}$_n$ (**20**) were obtained under exactly the same conditions, except using a slightly different molar ratio between CoCl$_2$·6H_2O and H_3btc (1.5:1 for **19** and 1.77:1 for **20**). However, these products feature very different structures and topologies (Figure 5). The structures of product pairs **21**/**22** and **23**/**24** (Table 3) also differ significantly on varying the NaOH:H$_2$cppa and NaOH:H$_3$bptc molar ratios, respectively. In these cases, an excess of sodium hydroxide leads to a complete deprotonation of H$_2$cppa in **22** or a generation of additional μ_3-OH linkers in **24**, thus making these structures more complicated in comparison with their counterparts assembled using a lower amount of NaOH.

Table 3. Selected examples of CPs showing an effect of reagents molar ratio on product structure.

Compound	Formula	Molar Ratio	Structure	Reference
19	{[Co$_3$(μ_4-btc)$_2$(μ-H$_2$O)$_2$(py)$_4$(H$_2$O)$_2$]·(py)$_2$}$_n$	CoCl$_2$:H$_3$btc = 1.5:1	3D MOF	[50]
20	{[Co$_{3.5}$(μ_6-btc)$_2$(μ_3-OH)(py)$_2$(H$_2$O)$_3$]·H$_2$O}$_n$	CoCl$_2$:H$_3$btc = 1.77:1	3D MOF	[50]
21	[Ni(Hcppa)$_2$(H$_2$O)$_2$]·2H$_2$O	NaOH:H$_2$cppa = 1:1	0D monomer	[44]
22	[Ni(μ_3-cppa)(H$_2$O)$_2$]$_n$	NaOH:H$_2$cppa = 2:1	2D layer	[44]
23	{[Zn$_3$(μ_6-bptc)$_2$(H$_2$O)$_4$]·H$_2$O}$_n$	NaOH:H$_3$bptc = 3:1	3D MOF	[51]
24	[Zn$_5$(μ_3-OH)$_4$(μ_6-bptc)$_2$(H$_2$O)$_2$]$_n$	NaOH:H$_3$bptc = 5:1	3D MOF	[51]

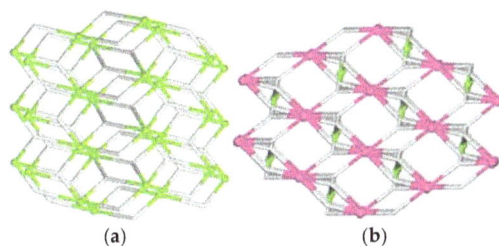

(a)   (b)

Figure 5. Topological representation of underlying 3D nets: (**a**) **ant** (anatase) net in **19**; (**b**) topologically unique net in **20** with the point symbol of $(4^2.6)_4(4^2.8^4)(4^6.6^4.8^{14}.10^4)$. Adapted from [50].

2.5. Effect of Reaction Temperature

The reaction temperature during the synthesis of metal-organic networks also has a significant impact on the final product structure. As illustrated in Table 4, compounds {[Co$_2$(μ_3-pyip)$_2$(DMF)]·(solv)}$_n$ (**25**) and {[Co(μ_3-pyip)]·2DMF}$_n$ (**26**) were synthesized from exactly the same reaction mixtures but at different temperatures, 80 and 120 °C, respectively. These 3D MOFs feature distinct structures (Figure 6).

Table 4. Selected examples of CPs showing an effect of reaction temperature on product structure.

Compound	Formula	Temperature (°C)	Structure	Reference
25	{[Co$_2$(μ_3-pyip)$_2$(DMF)]·(solv)}$_n$	80	3D MOF	[52]
26	{[Co(μ_3-pyip)]·2DMF}$_n$	120	3D MOF	[52]

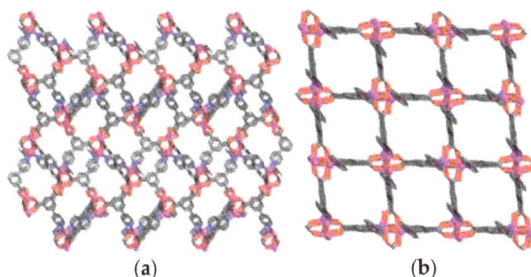

(a)   (b)

Figure 6. 3D metal-organic frameworks of **25** (**a**) and **26** (**b**). Adapted from [52].

2.6. Effect of Auxiliary Ligand

The presence of an additional auxiliary ligand also plays an important role in the hydrothermal synthesis of CPs, especially by facilitating product crystallization. Introduction of a common auxiliary *N,N*-donor ligand such as 2,2′-bipyridine of 1,10-phenanthroline usually changes the coordination environment of metal centers, thus resulting in the generation of different structures (Table 5). For example, the reaction of a cobalt(II) salt with H$_2$cppa with no auxiliary ligand leads to a 2D coordination polymer [Co(μ_3-cppa)(H$_2$O)$_2$]$_n$ (**27**), whereas simpler 1D zigzag chain products {[Co(μ-cppa)(2,2′-bpy)(H$_2$O)]·H$_2$O}$_n$ (**28**) and [Co(μ-cppa)(phen)(H$_2$O)]$_n$ (**29**) are generated in the presence of 2,2′-bpy or phen, respectively. Similarly, structurally distinct CPs {[Nd(μ-Hcpna)$_2$(μ-cpna)$_2$(H$_2$O)$_2$]·3H$_2$O}$_n$ (**34**) and {[Nd(μ-Hcpna)$_2$(μ_4-cpna)$_2$(phen)]·2H$_2$O}$_n$ (**35**) (Figure 7) were prepared under the same synthetic conditions except the introduction of phen in **35**. As can be seen from various examples collected in Table 5, the use of the *N,N*-donor auxiliary ligands tends to facilitate the formation of CPs with a lower dimensionality if compared to the systems without an auxiliary ligand. However, rather complex 3D MOF {[Cd$_3$(μ_5-btc)$_2$(phen)$_2$(H$_2$O)]·H$_2$O}$_n$ (**31**) can also be generated in the presence of the auxiliary ligand (Table 5).

Table 5. Selected examples of CPs showing an effect of auxiliary ligand on product structure.

Compound	Formula	Auxiliary Ligand	Structure	Reference
27	[Co(μ_3-cppa)(H$_2$O)$_2$]$_n$	no	2D network	[44]
28	{[Co(μ-cppa)(2,2′-bpy)(H$_2$O)]·H$_2$O}$_n$	2,2′-bpy	1D zigzag chain	[44]
29	[Co(μ-cppa)(phen)(H$_2$O)]$_n$	phen	1D zigzag chain	[44]
30	{[Cd$_3$(μ_6-btc)$_2$(H$_2$O)$_5$]·4H$_2$O}$_n$	no	3D MOF	[49]
31	{[Cd$_3$(μ_5-btc)$_2$(phen)$_2$(H$_2$O)]·H$_2$O}$_n$	phen	3D MOF	[49]
32	[Mn(μ_3-cpna)(2,2′-bpy)(H$_2$O)]$_n$	2,2′-bpy	2D layer	[42]
33	[Mn(μ_3-cpna)(phen)(H$_2$O)]$_n$	phen	1D ladder chain	[42]
34	{[Nd(μ-Hcpna)$_2$(μ-cpna)$_2$(H$_2$O)$_2$]·3H$_2$O}$_n$	no	2D layer	[42]
35	{[Nd(μ-Hcpna)$_2$(μ_4-cpna)$_2$(phen)]·2H$_2$O}$_n$	phen	1D double chain	[42]

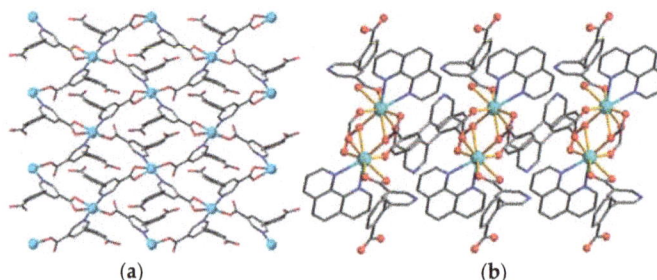

(a)   (b)

Figure 7. (a) 2D metal-organic layer in **34**. (b) 1D double chain in **35**. Adapted from [42].

2.7. Effect of Template

Template-assisted synthesis of CPs has attracted a special attention as a promising approach toward tunable architectures or structures that might be difficult to access by routine synthetic methods [47,53,54]. Various inorganic ions or organic molecules can be used as templating agents in the hydrothermal synthesis of coordination polymers. In particular, 4,4′-bipyridine acts not only as a common linker in CPs but is frequently applied as a template. Selected pairs of structurally distinct coordination polymers obtained with or without template are summarized in Table 6. For example, although compounds {[Ni$_3$(μ_4-dcppa)$_2$(H$_2$O)$_6$]·2H$_2$O}$_n$ (**42**) and {[Ni$_3$(μ_5-dcppa)$_2$(H$_2$O)$_6$]·2H$_2$O}$_n$ (**43**) were prepared under similar reaction conditions except using 4,4′-bipy as a templating agent in **43**, they feature structures of different dimensionality and topology (Figure 8).

Table 6. Selected examples of CPs showing an effect of template on product structure.

Compound	Formula	Template	Structure	Reference
36	$\{[Mn_2(\mu_3\text{-pyip})_2(H_2O)_4]\cdot 5H_2O\}_n$	no	2D layer	[55]
37	$[Mn_3(\mu_5\text{-pyip})_2(\mu\text{-HCOO})_2(H_2O)_2]_n$	4,4'-bpy	2D layer	[55]
38	$[Co(\mu_3\text{-pyip})(EtOH)(H_2O)]_n$	no	2D layer	[55]
39	$\{[Co(\mu_4\text{-pyip})(H_2O)]\cdot H_2O\}_n$	cyanoacetic acid	2D double layer	[55]
40	$\{[Mn_3(\mu_4\text{-dcppa})_2(H_2O)_6]\cdot 3H_2O\}_n$	no	2D layer	[47]
41	$\{[Mn_3(\mu_5\text{-dcppa})_2(H_2O)_6]\cdot 4H_2O\}_n$	4,4'-bpy	3D MOF	[47]
42	$\{[Ni_3(\mu_4\text{-dcppa})_2(H_2O)_6]\cdot 2H_2O\}_n$	no	2D layer	[47]
43	$\{[Ni_3(\mu_5\text{-dcppa})_2(H_2O)_6]\cdot 2H_2O\}_n$	4,4'-bpy	3D MOF	[47]

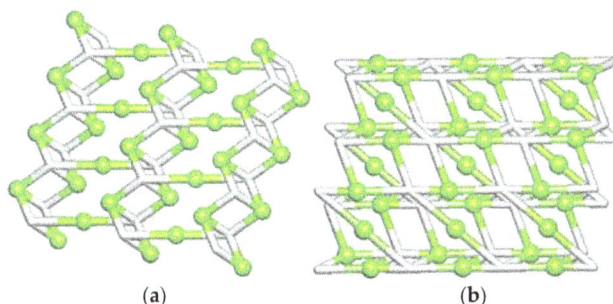

(a) (b)

Figure 8. Topological representation of underlying nets: (**a**) 2D layer with **3,4L83** topology in **42**; (**b**) 3D framework with **tcs** topology in **43**. Adapted from [47].

2.8. Effect of Two Main Ligands

Although a substantial number of coordination polymers incorporating various kinds of carboxylate ligands has been reported [56], the examples of heteroleptic networks constructed from a combination of two kinds of biphenyl or phenyl-pyridine carboxylate building blocks (Scheme 1) are barely known. It is primarily caused by different solubility of such ligands, distinct coordination modes and charges, as well as ligand competition for metal node during the hydrothermal synthesis and crystallization. The latter factor may often lead to the formation of a mixture of simpler products containing only one main building block rather than more complex products comprising both carboxylate ligands. The competition between two main carboxylate building blocks for metal nodes can be even more pronounced when the reaction mixture also contains an additional auxiliary ligand along with water as a solvent and frequent terminal ligand source. The effect of two different types of biphenyl carboxylate moieties on the structure of the resulting metal-organic network remains poorly studied. Notable examples of CPs combining two kinds of biphenyl carboxylate blocks include a 2D network $[Cd_2(\mu_5\text{-cpic})_2(\mu\text{-bpdc})_{0.5}(phen)_2]_n$ (**45**) and a 3D MOF $[Co_2(\mu_7\text{-btc})_2(\mu\text{-bpydc})_{0.5}(py)_3]_n$ (**47**) that feature distinct structures and topologies in comparison with their counterparts $\{[Cd_2(\mu_4\text{-cpic})(\mu_3\text{-OH})(phen)_2]\cdot 2H_2O\}_n$ (**44**) and $\{[Co_3(\mu_4\text{-btc})_2(\mu\text{-H}_2O)_2(py)_4(H_2O)_2]\cdot(py)_2\}_n$ (**46**), respectively (Table 7, Figure 9).

Table 7. Selected examples of CPs showing an effect of two main carboxylate ligands on product structure.

Compound	Formula	Main Ligand	Structure	Reference
44	$\{[Cd_2(\mu_4\text{-cpic})(\mu_3\text{-OH})(phen)_2]\cdot 2H_2O\}_n$	H_3cpic	2D layer	[57]
45	$[Cd_2(\mu_5\text{-cpic})_2(\mu\text{-bpdc})_{0.5}(phen)_2]_n$	H_3cpic, H_2bpdc	2D layer	[57]
46	$\{[Co_3(\mu_4\text{-btc})_2(\mu\text{-H}_2O)_2(py)_4(H_2O)_2]\cdot(py)_2\}_n$	H_3btc	3D MOF	[50]
47	$[Co_2(\mu_7\text{-btc})_2(\mu\text{-bpydc})_{0.5}(py)_3]_n$	H_3btc, H_2bpydc	3D MOF	[58]

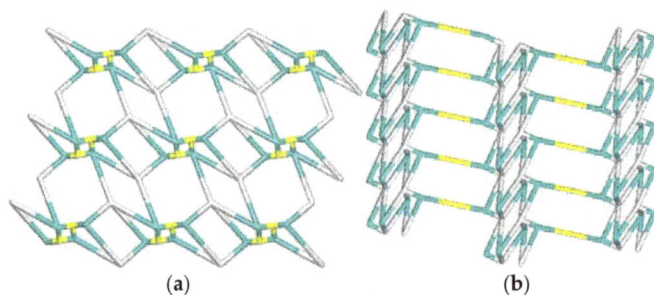

Figure 9. Topological representation of underlying nets: (**a**) 2D layer with **3,4L33** topology in **44**; (**b**) trinodal 3,3,5-connected 2D layer in **45** with the unique topology and point symbol of $(4.6^2)(4^3)(4^4.6^4.8^2)$. Adapted from [57].

3. Selected Functional Properties and Applications

3.1. Highly Porous MOFs

Some coordination polymers based on multifunctional carboxylic acids with phenyl-pyridine or biphenyl cores possess the highly porous structures and excellent stability (Table 8). These properties make these materials rather promising for exploring CO_2 capture and gas storage applications. As illustrated in Table 8 and Figure 10, Zhao an co-workers synthesized a UiO type MOF derived from the H_2bpydc block, $[Zr_6(\mu_3\text{-O})_4(OH)_4(\mu\text{-bpydc})_{12}]$ (**50**). This MOF exhibits high storage capacity for H_2, CH_4, and CO_2, showing an unusual stepwise adsorption for liquid CO_2 and solvents with a sequential filling mechanism on different adsorption sites. Other related MOFs with high porosity and interesting N_2, H_2, CO_2 and/or CH_4 uptake behavior include $[Cu_2(\mu_3\text{-pyip})_2(H_2O)_2]_{0.5}[Cu(pyip)]$ (**48**), $\{[Cu(\mu_3\text{-pyip})(H_2O)_2]\cdot1.5DMF\}_n$ (**49**), and $[Zn_3(\mu_5\text{-bpydc})_2(HCOO)_2]\cdot H_2O\cdot DMF$ (**51**) (Table 8).

Table 8. Selected examples of highly porous metal-organic frameworks (MOFs).

Compound	Formula	Porosity	Applications in Gas Uptake or Separation	Reference
48	$[Cu_2(\mu_3\text{-pyip})_2(H_2O)_2]_{0.5}[Cu(pyip)]$	60.8%	N_2, H_2, CO_2	[59]
49	$\{[Cu(\mu_3\text{-pyip})]\cdot2H_2O\cdot1.5DMF\}_n$	54.0%	N_2, H_2, CO_2	[60]
50	$[Zr_6(\mu_3\text{-O})_4(OH)_4(\mu\text{-bpydc})_{12}]$	68.5%	N_2, H_2, CO_2, CH_4	[61]
51	$[Zn_3(\mu_5\text{-bpydc})_2(HCOO)_2]\cdot H_2O\cdot DMF$	64.3%	N_2, CO_2, CH_4	[62]

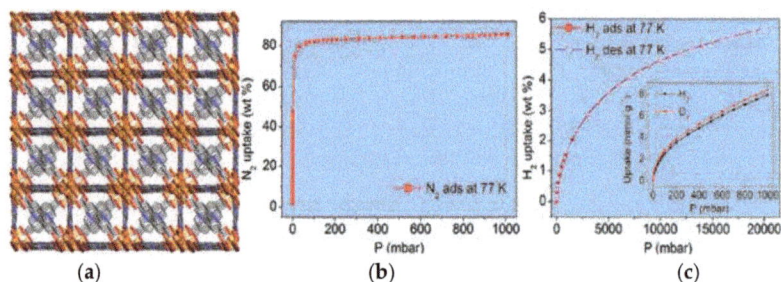

(a)   (b)   (c)

Figure 10. *Cont.*

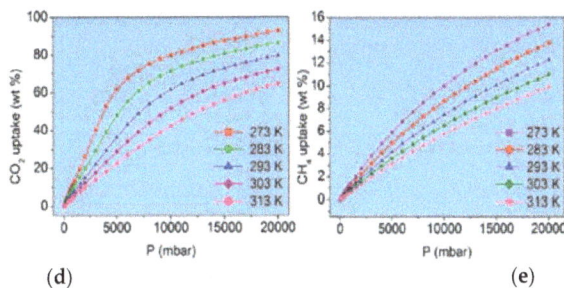

Figure 10. (a) 3D metal-organic framework of **50**. (**b–e**) Adsorption isotherms of **50** for (**b**) N_2, (**c**) H_2 and D_2 (inset), (**d**) CO_2, and (**e**) CH_4. Adapted from [61].

3.2. Highly Luminescent Materials

MOFs based on the europium(III) and terbium(III) nodes are highly luminescent compounds. As illustrated in Table 9 and Figure 11, an interesting example concerns a Tb MOF $[Tb(\mu_4\text{-bpydc})(\mu_3\text{-HCOO})]_n$ (**53**) derived from the H_2bpydc building block. It features a remarkable temperature-dependent photoluminescence. At 298 K, under UV excitation, compound **53** glows red-orange, whereas at 77 K it emits a green light. Another example concerns a Eu(III) derivative $[Eu_2(\mu_4\text{-pyip})_3(H_2O)_4]_n \cdot 2n\text{DMF} \cdot 3n\text{H}_2\text{O}$ (**52**) that is capable of emitting different colors ranging from yellow to red and orange.

Table 9. Selected examples of highly luminescent MOFs.

Compound	Formula	λ_{em} (nm)	Color	Reference
52	$[Eu_2(\mu_4\text{-pyip})_3(H_2O)_4]_n \cdot 2n\text{DMF} \cdot 3n\text{H}_2\text{O}$	255–365	yellow to red and then to orange	[63]
53	$[Tb(\mu_4\text{-bpydc})(\mu_3\text{-HCOO})]_n$	614, 541	red-orange (298 K), green (77 K)	[64]

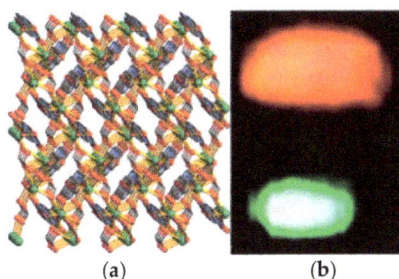

Figure 11. (**a**) 3D metal-organic framework of **53**; (**b**) temperature-dependent red-orange (top, 298 K) or green (bottom, 77 K) emission under UV excitation. Adapted from [64].

3.3. Compounds with Unusual Magnetic Properties

Some coordination polymers derived from multifunctional carboxylic acids with phenyl-pyridine or biphenyl cores can exhibit unusual magnetic properties. Selected examples are highlighted in Table 10. In particular, Du and co-workers assembled a 3D MOF, $\{[Dy_2(\mu_4\text{-pyip})_3(H_2O)_4] \cdot 2\text{DMF} \cdot 3\text{H}_2\text{O}\}_n$ (**54**), using H_2pyip as a building block. This compound possesses the **pcu** topology and exhibits a slow magnetization relaxation behavior (Figure 12). Other notable examples of magnetic CPs include a nickel(II) derivative $[Ni_3(\mu_5\text{-pyip})_2(\mu\text{-HCOO})_2(H_2O)_2]_n$ (**55**) with a long-range magnetic ordering as well as the dysprosium(III)

$[Dy(\mu_5\text{-bptc})(phen)(H_2O)]_n$ (**56**) and $\{[Dy_3Co_2(\mu_4\text{-bpydc})_5(\mu_3\text{-Hbpydc})(H_2O)_5](ClO_4)_2\}_n$ (**57**) frameworks with a slow magnetization relaxation behavior.

Table 10. Selected examples of CPs with unusual magnetic properties.

Compound	Formula	Magnetic Behavior	Highlight	Reference
54	$\{[Dy_2(\mu_4\text{-pyip})_3(H_2O)_4]\cdot 2DMF\cdot 3H_2O\}_n$	weak ferromagnetic	slow magnetization relaxation behavior	[63]
55	$[Ni_3(\mu_5\text{-pyip})_2(\mu\text{-HCOO})_2(H_2O)_2]_n$	weak ferromagnetic	long-range magnetic ordering	[65]
56	$[Dy(\mu_5\text{-bptc})(phen)(H_2O)]_n$	antiferromagnetic	slow magnetization relaxation behavior	[66]
57	$\{[Dy_3Co_2(\mu_4\text{-bpydc})_5(\mu_3\text{-Hbpydc}) (H_2O)_5](ClO_4)_2\cdot 11H_2O\}_n$	antiferromagnetic	slow magnetization relaxation behavior	[67]

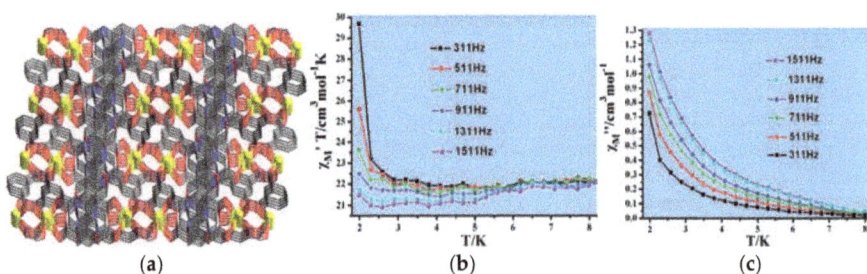

Figure 12. (**a**) 3D metal-organic framework of **54**. (**b,c**) Ac susceptibility of **54** measured in zero dc fields and plotted as $\chi'T$ vs. T (**b**) and χ'' vs. T (**c**). Adapted from [63].

3.4. Selective Sensing Materials

It is known that some fluorescent MOF materials are sensitive to the presence or absence of guest solvent molecules. As illustrated in Table 11 and Figure 13, Wen and co-workers reported a 3D MOF based on the H_2pyip ligand, $[Zn(\mu_3\text{-pyip})(bimb)\cdot(H_2O)]_n$ (**58**). This MOF exhibits the first report of a MOF material as a promising luminescent probe for detecting pesticides. This compound is also unique by allowing a detection of both pesticides and solvent molecules simultaneously. Other examples of sensing MOFs are shown in Table 11.

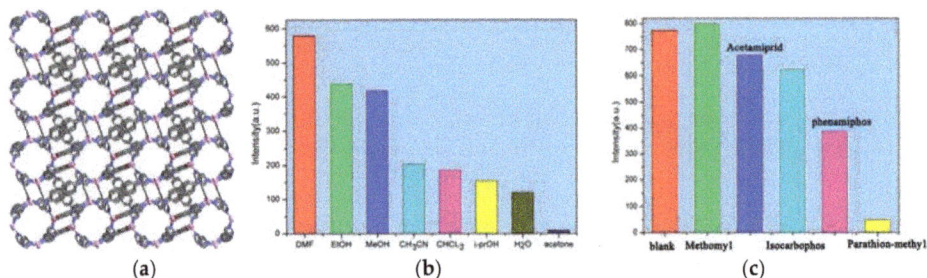

Figure 13. (**a**) 3D metal-organic framework of **58**. (**b,c**) Photoluminescence intensities of **58** introduced to (**b**) various pure solvents or (**c**) different pesticides (1×10^{-3} M in DMF); λ_{ex} = 290 nm. Adapted from [68].

Table 11. Selected examples of MOFs with selective sensing behavior.

Compound	Formula	Structure	Analyte	Reference
58	$[Zn(\mu_3\text{-pyip})(bimb)\cdot(H_2O)]_n$	3D MOF	acetone, pesticides	[68]
59	$[Zr_6(\mu_3\text{-O})_4(OH)_4(\mu_4\text{-bpydc})_{12}]_n$	3D MOF	Fe^{3+} ions	[69]
60	$[Eu_2(\mu_4\text{-bpydc})_3(H_2O)_3]_n\cdot nDMF$	3D MOF	Cu^{2+} ions	[69]

4. Conclusions and Outlook

In this mini-review, we featured selected recent examples of coordination polymers (CPs) or metal-organic frameworks (MOFs) that were constructed from various multifunctional carboxylic acids with phenyl-pyridine or biphenyl cores (Scheme 1). Despite being still little explored, these types of semi-rigid, thermally stable, and versatile building blocks appear to be very promising for the hydrothermal synthesis of metal-organic networks with different structural characteristics, topologies, and functional properties. The present work also highlighted an importance of different reaction parameters and conditions on the assembly and structural diversity of coordination polymers. The effects of the type of main carboxylate ligand, kind of metal node, stoichiometry and molar ratio of reagents, temperature, presence or absence of auxiliary ligands or templates were showcased. In addition, some examples of highly porous MOFs, notable luminescent materials, compounds with unusual magnetic properties, and frameworks for selective sensing applications were described.

We believe the application of multifunctional carboxylic acids containing phenyl-pyridine or biphenyl cores toward the design of coordination polymers will be continued, leading to new series of coordination compounds and derived materials with fascinating structural features and notable functional properties. Future research might focus on: (A) widening the family of multicarboxylate building blocks to new members with additional functional groups; (B) diversifying the types of metal nodes; (C) assembling heterometallic metal-organic architectures; (D) optimizing the conditions of the hydrothermal synthesis and crystallization; (E) predicting the structural and topological characteristics; and (F) broadening the types of possible applications of the obtained coordination polymers.

Acknowledgments: This work was financially supported by the National Natural Science Foundation of China (Project 21572086). AMK and MVK acknowledge the Foundation for Science and Technology (FCT), Portugal (UID/QUI/00100/2013, IF/01395/2013/CP1163/CT005).

Conflicts of Interest: The authors declare no conflict of interest.

Abbreviations

0D	zero-dimensional
1D	one-dimensional
2D	two-dimensional
3D	three-dimensional
CP	coordination polymer
MOF	metal-organic framework
H_2cpna	5-(2′-carboxylphenyl)-nicotinic acid
H_2pyip	5-(4-pyridyl)-isophthalic acid
H_2cppa	4-(3-carboxyphenyl)-picolinic acid
H_2bpydc	2,2′-bipyridine-5,5′-dicarboxylic acid
H_3bptc	biphenyl-2,5,3′-tricarboxylic acid
H_3btc	biphenyl-2,4,4′-tricarboxylic acid
H_3cpic	4-(5-carboxypyridin-2-yl)-isophthalic acid
H_3cptc	2-(4-carboxypyridin-3-yl)-terephthalic acid
H_3dcppa	5-(6-carboxypyridin-3-yl)-isophthalic acid
H_3cpta	2-(5-carboxypyridin-2-yl)-terephthalic acid
py	pyridine
phen	1,10-phenanthroline

2,2′-bpy	2,2′-bipyridine
4,4′-bpy	4,4′-bipyridine
H_2biim	2,2′-biimidazole
H_2bpdc	4,4′-biphenyldicarboxylic acid
bimb	4,4′-bis(1-imidazolyl)biphenyl

References

1. Batten, S.R.; Champness, N.R.; Chen, X.M.; Garcia-Martinez, J.; Kitagawa, S.; Öhrström, L.; O'Keeffe, M.; Suh, M.P.; Reedijk, J. Terminology of metal-organic frameworks and coordination polymers (IUPAC Recommendations 2013). *Pure Appl. Chem.* **2013**, *85*, 1715–1724. [CrossRef]
2. Kaskel, S. (Ed.) *The Chemistry of Metal-Organic Frameworks: Synthesis, Characterization, and Applications*; Wiley: Hoboken, NJ, USA, 2016; p. 904.
3. Banerjee, R. (Ed.) *Functional Supramolecular Materials: From Surfaces to MOFs*; RSC: London, UK, 2017; p. 461.
4. MacGillivray, L.R.; Lukehart, C.M. (Eds.) *Metal-Organic Framework Materials*; Wiley: Hoboken, NJ, USA, 2014; p. 592.
5. Farrusseng, D. (Ed.) *Metal-Organic Frameworks: Applications from Catalysis to Gas Storage*; Wiley: Hoboken, NJ, USA, 2011; p. 414.
6. Batten, S.R.; Neville, S.M.; Turner, D.R. *Coordination Polymers: Design, Analysis and Application*; RSC: London, UK, 2009; p. 424.
7. Schmidt, F.M.S.; Merkel, M.P.; Kostakis, G.E.; Buth, G.; Anson, C.E.; Powell, A.K. SMM behaviour and magnetocaloric effect in heterometallic 3d-4f coordination clusters with high azide: Metal ratios. *Dalton Trans.* **2017**, *46*, 15661–15665. [CrossRef] [PubMed]
8. Kallitsakis, M.; Loukopoulos, E.; Abdul-Sada, A.; Tizzard, G.J.; Coles, S.J.; Kostakis, G.E.; Lykakis, I.N. A copper-benzotriazole-based coordination polymer catalyzes the efficient one-pot synthesis of (*N*″-substituted)-hydrazo-4-aryl-1,4-dihydropyridines from Azines. *Adv. Synth. Catal.* **2017**, *359*, 138–145. [CrossRef]
9. Catala, L.; Mallah, T. Nanoparticles of prussian blue analogs and related coordination polymers: From information storage to biomedical applications. *Coord. Chem. Rev.* **2017**, *346*, 32–61. [CrossRef]
10. Islamoglu, T.; Goswami, S.; Li, Z.Y.; Howarth, A.J.; Farha, O.K.; Hupp, J.T. Postsynthetic tuning of metal-organic frameworks for targeted applications. *Acc. Chem. Res.* **2017**, *50*, 805–813. [CrossRef] [PubMed]
11. Lusby, P.J. Supramolecular coordination chemistry. *Annu. Rep. Prog. Chem. Sect. A* **2010**, *106*, 319–339. [CrossRef]
12. Liu, S.J.; Chen, Y.; Xu, W.J.; Zhao, Q.; Huang, W. New trends in the optical and electronic applications of polymers containing transition-metal complexes. *Macromol. Rapid Commun.* **2012**, *33*, 461–480. [CrossRef] [PubMed]
13. Kaur1, R.; Kim, K.H.; Paul, A.K.; Akash, D. Recent advances in the photovoltaic applications of coordination polymers and metal organic frameworks. *J. Mater. Chem. A* **2016**, *4*, 3991–4002. [CrossRef]
14. He, C.B.; Liu, D.M.; Lin, W.B. Nanomedicine applications of hybrid nanomaterials built from metal-ligand coordination bonds: Nanoscale metal-organic frameworks and nanoscale coordination polymers. *Chem. Rev.* **2015**, *115*, 11079–11108. [CrossRef] [PubMed]
15. Mai, H.D.; Rafiq, K.; Yoo, H. Nano metal-organic framework-derived inorganic hybrid nanomaterials: Synthetic strategies and applications. *Chem. Eur. J.* **2017**, *23*, 5631–5651. [CrossRef] [PubMed]
16. Cui, Y.J.; Li, B.; He, H.J.; Zhou, W.; Chen, B.L.; Qian, G.D. Metal-organic frameworks as platforms for functional materials. *Acc. Chem. Res.* **2016**, *49*, 483–493. [CrossRef] [PubMed]
17. Dolgopolova, E.A.; Brandt, A.J.; Ejegbavwo, O.A.; Duke, A.S.; Maddumapatabandi, T.D.; Galhenage, R.P.; Larson, B.W.; Reid, O.G.; Ammal, S.C.; Heyden, A.; et al. Electronic properties of bimetallic metal-organic frameworks (MOFs): Tailoring the density of electronic states through MOF modularity. *J. Am. Chem. Soc.* **2017**, *139*, 5201–5209. [CrossRef] [PubMed]
18. Kirillov, A.M.; Wieczorek, S.W.; Lis, A.; da Silva, M.F.C.G.; Florek, M.; Król, J.; Staroniewicz, Z.; Smolenski, P.; Pombeiro, A.J.L. 1,3,5-Triaza-7-phosphaadamantane-7-oxide (PTA=O): New diamondoid building block for design of 3D metal-organic frameworks. *Cryst. Growth Des.* **2011**, *11*, 2711–2716. [CrossRef]

19. Dias, S.S.P.; Kirillova, M.V.; André, V.; Kłak, J.; Kirillov, A.M. New tetracopper(II) cubane cores driven by a diamino alcohol: Self-assembly synthesis, structural and topological features, and magnetic and catalytic oxidation properties. *Inorg. Chem.* **2015**, *54*, 5204–5212. [CrossRef] [PubMed]

20. Lemaire, P.C.; Lee, D.T.; Zhao, J.J.; Parsons, G.N. Reversible low-temperature metal node distortionduring atomic layer deposition of Al_2O_3 and TiO_2 onUiO-66-NH_2metal organic framework crystal surfaces. *ACS Appl. Mater. Interfaces* **2017**, *9*, 22042–22054. [CrossRef] [PubMed]

21. Ji, P.F.; Manna, K.; Lin, Z.; Urban, A.; Greene, F.X.; Lan, G.X.; Lin, W.B. Single-site cobalt catalysts at new $Zr_8(\mu_2\text{-O})_8(\mu_2\text{-OH})_4$metal-organic framework nodes for highly active hydrogenation of alkenes, imines, carbonyls, and heterocycles. *J. Am. Chem. Soc.* **2016**, *138*, 12234–12242. [CrossRef] [PubMed]

22. Jaros, S.W.; Silva, M.F.C.G.D.; Florek, M.; Oliveira, M.C.; Smoleński, P.; Pombeiro, A.J.L.; Kirillov, A.M. Aliphatic dicarboxylate directed assembly of silver(I) 1,3,5-Triaza-7-phosphaadamantane coordination networks: Topological versatility and antimicrobial activity. *Cryst. Growth Des.* **2014**, *14*, 5408–5417. [CrossRef]

23. Wang, K.B.; Geng, Z.R.; Zheng, M.B.; Ma, L.Y.; Ma, X.Y.; Wang, Z.L. Controllable fabrication of coordination polymer particles (CPPs): A bridge between versatile organic building blocks and porous copper-based inorganic materials. *Cryst. Growth Des.* **2012**, *12*, 5606–5614. [CrossRef]

24. Manna, P.; Das, S.K. Perceptive approach in assessing rigidity versus flexibility in the construction of diverse metal-organic coordination networks: Synthesis, structure, and magnetism. *Cryst. Growth Des.* **2015**, *15*, 1407–1421. [CrossRef]

25. Beeching, L.J.; Hawes, C.S.; Turnera, D.R.; Batten, S.R. The influence of anion, ligand geometry and stoichiometry on the structure and dimensionality of a series of AgI-bis(cyanobenzyl)piperazine coordination polymers. *CrystEngComm* **2014**, *16*, 6459–6468. [CrossRef]

26. Xu, W.; Si, Z.X.; Xie, M.; Zhou, L.X.; Zheng, Y.Q. Experimental and theoretical approaches to three uranyl coordination polymers constructed by phthalic acid and N,N′-donor bridging ligands: crystal structures, luminescence, and photocatalytic degradation of tetracycline hydrochloride. *Cryst. Growth Des.* **2017**, *17*, 2147–2157. [CrossRef]

27. Kobayashi, Y.; Honjo, K.; Kitagawa, S.; Uemura, T. Preparation of porous polysaccharides templated by coordination polymer with three-dimensional nanochannels. *ACS Appl. Mater. Interfaces* **2017**, *9*, 11373–11379. [CrossRef] [PubMed]

28. Ding, R.; Huang, C.; Lu, J.J.; Wang, J.N.; Song, C.J.; Wu, J.; Hou, H.W.; Fan, Y.T. Solvent templates induced porous metal-organic materials: Conformational isomerism and catalytic activity. *Inorg. Chem.* **2015**, *54*, 1405–1413. [CrossRef] [PubMed]

29. Garai, M.; Maji, K.; Chernyshev, V.V.; Biradha, K. Interplay of pyridine substitution and Ag(I)···Ag(I) and Ag(I)···π interactions in templating photochemical solid state [2+2] reactions of unsymmetrical olefins containing amides: Single-crystal-to-single-crystal transformations of coordination polymers. *Cryst. Growth Des.* **2016**, *16*, 550–554. [CrossRef]

30. Wang, J.; Bai, C.; Hua, H.M.; Yuan, F.; Xue, G.L. A family of entangled coordination polymers constructed from a flexible V-shaped long bicarboxylic acid and auxiliary N-donor ligands: Luminescent sensing. *J. Solid State Chem.* **2017**, *249*, 87–97. [CrossRef]

31. Zhang, J.Y.; Shi, J.X.; Chen, L.Y.; Jia, Q.X.; Deng, W.; Gao, E.Q. N-donor auxiliary ligand-directed assembly of Co^{II} compounds with a 2,2′-dinitro-biphenyl-4,4′-dicarboxylate ligand: Structures and magnetic properties. *CrystEngComm* **2017**, *19*, 1738–1750. [CrossRef]

32. Ren, Y.L.; Li, L.; Mu, B.; Li, C.X.; Huang, R.D. Electrocatalytic properties of three new POMs-based inorganic-organic frameworks with flexible zwitterionic dicarboxylate ligands. *J. Solid State Chem.* **2017**, *249*, 1–8. [CrossRef]

33. Hu, K.Q.; Zhu, L.Z.; Wang, C.Z.; Mei, L.; Liu, Y.H.; Gao, Z.Q.; Chai, Z.F.; Shi, W.Q. Novel uranyl coordination polymers based on quinoline-containing dicarboxylate by altering auxiliary ligands: From 1D chain to 3D framework. *Cryst. Growth Des.* **2016**, *16*, 4886–4896. [CrossRef]

34. Yang, L.Z.; Wang, J.; Kirillov, A.M.; Dou, W.; Xu, C.; Fang, R.; Xu, C.L.; Liu, W.S. 2D lanthanide MOFs driven by a rigid 3,5-bis(3-carboxy-phenyl)pyridine building block: Solvothermal syntheses, structural features, and photoluminescence and sensing properties. *CrystEngComm* **2016**, *18*, 6425–6436. [CrossRef]

35. Song, J.F.; Jia, Y.Y.; Zhou, R.S.; Li, S.Z.; Qiu, X.M.; Liu, J. Six new coordination compounds based on rigid 5-(3-carboxy-phenyl)-pyridine-2-carboxylic acid: Synthesis, structural variations and properties. *RSC Adv.* **2017**, *7*, 7217–7226. [CrossRef]

36. Øien, S.; Agostini, G.; Svelle, S.; Borfecchia, E.; Lomachenko, K.A.; Mino, L.; Gallo, E.; Bordiga, S.; Olsbye, U.; Lillerud, K.P.; et al. Probing reactive platinum sites in UiO-67 zirconium metal-organicframeworks. *Chem. Mater.* **2015**, *27*, 1042–1056.

37. Amarante, T.R.; Neves, P.; Valente, A.A.; Paz, F.A.A.; Fitch, A.N.; Pillinger, M.; Gonçalves, I.S. Hydrothermal synthesis, crystal structure, and catalytic potential of a one-dimensional molybdenum oxide/bipyridinedicarboxylate hybrid. *Inorg. Chem.* **2013**, *52*, 4618–4628. [CrossRef] [PubMed]

38. Maza, W.A.; Padilla, R.; Morris, A.J. Concentration dependent dimensionality of resonance energy transfer in a postsynthetically doped morphologically homologous analogue of UiO-67 MOF with a ruthenium(II) polypyridyl complex. *J. Am. Chem. Soc.* **2015**, *137*, 8161–8168. [CrossRef] [PubMed]

39. Robin, A.Y.; Fromm, K.M. Coordination polymer networks with O- and N-donors: What they are, why and how they are made. *Coord. Chem. Rev.* **2006**, *250*, 2127–2157. [CrossRef]

40. Byrappa, K.; Yoshimura, M. *Handbook of Hydrothermal Technology: A Technology for Crystal Growth and Materials Processing*; William Andrew Publishing, LLC Norwich: New York, NY, USA, 2001.

41. Kitagawa, S.; Noro, S. Coordination polymers: Infinite systems. *Compr. Coord. Chem.* **2003**, *7*, 231–256.

42. Gu, J.Z.; Gao, Z.Q.; Tang, Y. pH and auxiliary ligand influence on the structural variations of 5(2′-Carboxylphenyl) nicotate coordination polymers. *Cryst. Growth Des.* **2012**, *12*, 3312–3323. [CrossRef]

43. Gu, J.Z.; Wu, J.; Lv, D.Y.; Tang, Y.; Zhu, K.Y.; Wu, J.C. Lanthanide coordination polymers based on 5-(2′-carboxylphenyl) nicotinate: Syntheses, structure diversity, dehydration/hydration, luminescence and magnetic properties. *Dalton Trans.* **2013**, *42*, 4822–4830. [CrossRef] [PubMed]

44. Gu, J.Z.; Liang, X.X.; Cui, Y.H.; Wu, J.; Kirillov, A.M. Exploring 4-(3-carboxyphenyl)picolinic acid as a semirigid building block for the hydrothermal self-assembly of diverse metal-organic and supramolecular networks. *CrystEngComm* **2017**, *19*, 117–118. [CrossRef]

45. Zhang, G.M.; Li, Y.; Zou, X.Z.; Zhang, J.A.; Gu, J.Z.; Kirillov, A.M. Nickel(II) and manganese(II) metal-organic networks driven by 2,2′bipyridine-5,5′dicarboxylate blocks: synthesis, structural features, and magnetic properties. *Transit. Met. Chem.* **2016**, *41*, 153–160. [CrossRef]

46. Shao, Y.L.; Cui, Y.H.; Gu, J.Z.; Kirillov, A.M.; Wu, J.; Wang, Y.W. A variety of metal-organic and supramolecular networks constructed from a new flexible multifunctional building block bearing picolinate and terephthalate functionalities: hydrothermal self-assembly, structural features, magnetic and luminescent properties. *RSC Adv.* **2015**, *5*, 87484–87495. [CrossRef]

47. Gu, J.Z.; Cui, Y.H.; Liang, X.X.; Wu, J.; Lv, D.Y.; Kirillov, A.M. Structurally distinct metal-organic and H–Bonded networks derived from 5-(6-Carboxypyridin-3-yl)isophthalic acid: Coordination and template effect of 4,4′-bipyridine. *Cryst. Growth Des.* **2016**, *16*, 4658–4670. [CrossRef]

48. Li, Y.; Zhou, Q.; Gu, J.Z.; You, A. Syntheses, crystal structures, and magnetic properties of Mn(II) and Co(II) coordination polymers constructed from pyridine-tricarboxylate ligand. *Chin. J. Struct. Chem.* **2017**, *36*, 661–670.

49. Gu, J.Z.; Kirillov, A.M.; Wu, J.; Lv, D.Y.; Tang, Y.; Wu, J.C. Synthesis, structural versatility, luminescent and magnetic properties of a series of coordination polymers constructed from biphenyl-2,4,4′-tricarboxylate and different N-donor ligands. *CrystEngComm* **2013**, *15*, 10287–10303. [CrossRef]

50. Shao, Y.L.; Cui, Y.H.; Gu, J.Z.; Wu, J.; Wang, Y.W.; Kirillov, A.M. Exploring biphenyl-2,4,4′-tricarboxylic acid as a flexible building block for the hydrothermal self-assembly of diverse metal-organic and supramolecular networks. *CrystEngComm* **2016**, *18*, 765–778. [CrossRef]

51. Wu, W.P.; Liu, B.; Yang, G.P.; Miao, H.H.; Xi, Z.P.; Wang, Y.Y. Two new pH-controlled coordination polymers constructed from an asymmetrical tricarboxylate ligand and Zn-based rod-shaped SBUs. *Inorg. Chem. Commun.* **2015**, *56*, 8–12. [CrossRef]

52. Jia, G.H.; Athwal, H.S.; Blake, A.J.; Champness, N.R.; Hubberstey, P.; Schroder, M. Increasing nuclearity of secondary building units in porous cobalt(II) metal-organic frameworks: Variation in structure and H_2 adsorption. *Dalton Trans.* **2011**, *40*, 12342–12349. [CrossRef] [PubMed]

53. Stock, N.; Biswas, S. Synthesis of metal-organic frameworks (MOFs): Routes to various MOF topologies, morphologies, and composites. *Chem. Rev.* **2012**, *112*, 933–969. [CrossRef] [PubMed]

54. Lu, Y.B.; Jian, F.M.; Jin, S.; Zhao, J.W.; Xie, Y.R.; Luo, G.T. Three-dimensional extended frameworks constructed from dinuclear lanthanide(III) 1,4-naphthalenedicarboxylate units with bis(2,2′-biimidazole) templates: Synthese, structures, and magnetic and luminescent properties. *Cryst. Growth Des.* **2014**, *14*, 1684–1694. [CrossRef]

55. Liu, Y.Y.; Li, H.J.; Han, Y.; Lv, X.F.; Hou, H.W.; Fan, Y.T. Template-assisted synthesis of Co, Mn-MOFs with magnetic properties based on pyridinedicarboxylic acid. *Cryst. Growth Des.* **2012**, *12*, 3505–3513. [CrossRef]

56. Groom, C.R.; Bruno, I.J.; Lightfoot, M.P.; Ward, S.C. Cambridge Structural Database. *Acta Cryst.* **2016**, *B72*, 171–179.

57. Gu, J.Z.; Cui, Y.H.; Wu, J.; Kirillov, A.M. A series of mixed-ligand 2D and 3D coordination polymers assembled from a novel multifunctional pyridine-tricarboxylate building block: Hydrothermal syntheses, structural and topological diversity, andmagnetic and luminescent properties. *RSC Adv.* **2015**, *5*, 78889–78901. [CrossRef]

58. Li, W.B.; Gao, Z.Q.; Gu, J.Z. Synthesis, crystal structure, and magnetic properties of a Co metal-organic framework with mixed dicarboxylate and tricarboxylate ligands. *Chin. J. Struct. Chem.* **2016**, *35*, 257–263.

59. Lin, Q.P.; Bu, X.H.; Feng, P.Y. Perfect statistical symmetrization of a heterofunctional ligand induced by pseudo-copper trimer in an expanded matrix of HKUST-1. *Cryst. Growth Des.* **2013**, *13*, 5175−5178. [CrossRef]

60. Xiang, S.L.; Huang, J.; Li, L.; Zhang, J.Y.; Jiang, L.; Kuang, X.J.; Su, C.Y. Nanotubular metal-organic frameworks with high porosity based on T-Shaped pyridyl dicarboxylate ligands. *Inorg. Chem.* **2011**, *50*, 1743–1748. [CrossRef] [PubMed]

61. Li, L.J.; Tang, S.F.; Wang, C.; Lv, X.X.; Jiang, M.; Wu, H.Z.; Zhao, X.B. High gas storage capacities and stepwise adsorption in a UiO type metal-organic framework incorporating Lewis basic bipyridyl sites. *Chem. Commun.* **2014**, *50*, 2304–2307. [CrossRef] [PubMed]

62. Wang, J.; Luo, J.H.; Zhao, J.; Li, D.S.; Li, G.H.; Huo, Q.S.; Liu, Y.L. Assembly of two flexible metal-organic frameworks with stepwise gas adsorption and highly selective CO_2 adsorption. *Cryst. Growth Des.* **2014**, *14*, 2375–2380. [CrossRef]

63. Li, Q.P.; Du, S.W. A family of 3D lanthanide organic frameworks with tunable luminescence and slow magnetic relaxation. *RSC Adv.* **2015**, *5*, 9898–9903. [CrossRef]

64. Min, Z.Y.; Singh-Wilmot, M.A.; Cahill, C.L.; Andrews, M.; Taylor, R. Isoreticular lanthanide metal-organic frameworks: Syntheses, structures and photoluminescence of a family of 3D phenylcarboxylates. *Eur. J. Inorg. Chem.* **2012**, 4419–4426. [CrossRef]

65. Meng, X.X.; Zhang, X.J.; Bing, Y.M.; Xu, N.; Shi, W.; Cheng, P. In situ generation of NiO nanoparticles in a magnetic metal-organic framework exhibiting three-dimensional magnetic ordering. *Inorg. Chem.* **2016**, *55*, 12938–12943. [CrossRef] [PubMed]

66. Zhao, J.; Zhu, G.H.; Xie, L.Q.; Wu, Y.S.; Wu, H.L.; Zhou, A.J.; Wu, Z.Y.; Wang, J.; Chen, Y.C.; Tong, M.L. Magnetic and luminescent properties of lanthanide coordination polymers with asymmetric biphenyl-3,2′,5′-tricarboxylate. *Dalton Trans.* **2015**, *44*, 14424–14435. [CrossRef] [PubMed]

67. Fang, M.; Shi, P.F.; Zhao, B.; Jiang, D.X.; Cheng, P.; Shi, W. A series of 3d-4f heterometallic three-dimensional coordination polymers: Syntheses, structures and magnetic properties. *Dalton Trans.* **2012**, *41*, 6820–6826. [CrossRef] [PubMed]

68. Zheng, X.F.; Zhou, L.; Huang, Y.M.; Wang, C.G.; Duan, J.G.; Wen, L.L.; Tian, Z.F.; Li, D.F. A series of metal-organic frameworks based on 5-(4-pyridyl)-isophthalic acid: selective sorption and fluorescence sensing. *J. Mater. Chem. A* **2014**, *2*, 12413–12422. [CrossRef]

69. Lin, X.P.; Hong, Y.H.; Zhang, C.; Huang, R.Y.; Wang, C.; Lin, W.B. Pre-concentration and energy transfer enable the efficient luminescence sensing of transition metal ions by metal-organic frameworks. *Chem. Commun.* **2015**, *51*, 16996–16999. [CrossRef] [PubMed]

MDPI

St. Alban-Anlage 66

4052 Basel, Switzerland

Tel. +41 61 683 77 34

Fax +41 61 302 89 18

http://www.mdpi.com

Crystals Editorial Office

E-mail: crystals@mdpi.com

http://www.mdpi.com/journal/crystals

www.ingramcontent.com/pod-product-compliance
Lightning Source LLC
Chambersburg PA
CBHW051911210326
41597CB00033B/6112